T0212224

Three-Dimensional Integration and Modeling: A Revolution in RF and Wireless Packaging

© Springer Nature Switzerland AG 2022

Reprint of original edition © Morgan & Claypool 2008

Three-Dimensional Integration and Modeling: A Revolution in RF and Wireless Packaging

Jong-Hoon Lee and Manos M. Tentzeris

ISBN: 978-3-031-00575-6   paperback
ISBN: 978-3-031-00575-6   paperback

ISBN: 978-3-031-01703-2   ebook
ISBN: 978-3-031-01703-2   ebook

DOI: 10.1007/978-3-031-01703-2

A Publication in the Springer series

SYNTHESIS LECTURES ON COMPUTATIONAL ELECTROMAGNETICS #17

Lecture #17
Series Editor: Constantine A. Balanis, Arizona State University

Series ISSN

ISSN 1932-1252   print
ISSN 1932-1716   electronic

Three-Dimensional Integration and Modeling: A Revolution in RF and Wireless Packaging

Jong-Hoon Lee
RFMD

Manos M. Tentzeris
Georgia Institute of Technology

SYNTHESIS LECTURES ON COMPUTATIONAL ELECTROMAGNETICS #17

ABSTRACT

This book presents a step-by-step discussion of the 3D integration approach for the development of compact system-on-package (SOP) front-ends. Various examples of fully-integrated passive building blocks (cavity/microstip filters, duplexers, antennas), as well as a multilayer ceramic (LTCC) V-band transceiver front-end midule demonstrate the revolutionary effects of this approach in RF/Wireless packaging and multifunctional miniaturization.

Designs covered are based on novel ideas and are presented for the first time for millimeter-wave (60GHz) ultrabroadband wireless modules.

KEYWORDS

Three Dimensional Integration, Bandpass Filter, Antenna, Low-Temperature Co-Fired Ceramic, Liquid Crystal Polymer organics, ceramics, soft surfaces, front-end modules, Front-End Module, Transceiver, Patch Resonator, Gigabit, Dual-Mode, Cavity, Millimeter Wave, V-band, 60 GHz

Contents

INTRODUCTION

In recent years, great advancements have been made in understanding the mechanisms of the functioning of the human brain. Technological developments such as functional magnetic resonance imaging (f MRI), positron emission tomography (PET), and magnetoencephalography (MEG) have made possible the mapping of the images of cerebral activity from hemodynamic, metabolic or electromagnetic measurements. Among these brain imaging techniques, electroencephalography (EEG) is unique in terms of simplicity, accessibility, and temporal resolution, and has been viewed with renewed interest in recent years, thanks to the use of advanced methods of analysis and interpretation of its data. These methods are able to improve the spatial resolution of conventional EEG, making it possible to address the analysis of the brain activity in a noninvasive way using the temporal resolution of brain phenomena (of the order of milliseconds). With high-resolution EEG, it is now possible to obtain cortical activation maps describing the activity of the brain at the cortical level during the execution of a given experimental task.

Simple imaging of regions of the brain activated during particular tasks does not, however, convey the information about how these regions communicate with each other for making the execution of the task possible. The concept of brain connectivity is viewed as central to the understanding of the organized behavior of cortical regions, beyond the simple mapping of their activities [1,2]. Such behavior is thought to be based on the interaction between different cortical sites and differently specialized ones. Cortical connectivity estimation aims to describe these interactions in connectivity patterns, which hold the direction and strength of the information flow between cortical areas. To this purpose, several methods have been developed and applied to data gathered from hemodynamic and electromagnetic techniques [3–7]. Two main definitions of brain connectivity have been proposed during recent years: *functional* and *effective* connectivity [8]. Functional connectivity is defined as the temporal correlation between spatially remote neurophysiologic events. Effective connectivity is defined as the simplest brain circuit which would produce the same temporal relationship between cortical sites as observed experimentally.

As for the functional connectivity, the methods proposed in literature typically involve estimation of some covariance properties between different time series. These properties are measured from different spatial sites during motor and cognitive tasks by EEG and f MRI techniques [4,5,7,9].

Structural equation modeling (SEM) is a technique that has been used recently to assess the connectivity between cortical areas in humans from hemodynamic and metabolic measurements [3,10–12]. The basic idea of SEM considers the covariance structure of the data [10]. The estimation of effective cortical connectivity obtained from f MRI data has, however, a low temporal resolution (of the order of seconds) which is far from the time scale in which the brain normally operates. Hence, it is interesting to know whether the SEM technique can be applied to cortical activity which are obtained by applying linear inverse techniques to high-resolution EEG data [5,13–15].

As important information in the EEG signals are coded in frequency domain rather than time domain (reviewed in [16]), attention was focused on detecting frequency-specific interactions in EEG or MEG signals; for instance, the coherence between the activity of pairs of channels [17–19]. However, coherence analysis does not have a directional nature (i.e., it just examines whether a link exists between two neural structures by describing instances when they are in synchronous activity) and does not provide directly the direction of the information flow. In this respect, multivariate spectral techniques such as directed transfer function (DTF) or partial directed coherence (PDC) were proposed [20,21] for determining the directional influences between any given pair of channels in a multivariate data set. Both DTF and PDC [21,22] rely on the key concept of Granger causality between time series [23], according to which an observed time series $x(n)$ causes another series $y(n)$ if the knowledge of $x(n)$'s past significantly improves the prediction of $y(n)$; this relation between time series is not reciprocal, i.e., $x(n)$ may cause $y(n)$ without $y(n)$ necessarily causing $x(n)$. This lack of reciprocity allows the evaluation of the direction of information flow between structures. These estimators are able to characterize at the same time both the directional and spectral properties of the brain signals, and they require only one multivariate autoregressive (MVAR) model estimated from all the EEG channels. The advantages of MVAR modeling of multichannel EEG signals were stressed recently [24] by demonstrating the advantages of multivariate methods with respect to the pairwise autoregressive approach, in terms of both accuracy and computational cost. In order to fully characterize the techniques presented we test them on simulated EEG data whose connectivity characteristics are known in advance, and then we finally apply such methods to human data obtained from high resolution EEG recordings. As a novelty, the application of all the proposed methodologies (SEM, DTF, and PDC) was performed using the cortical signals estimated from high-resolution EEG recordings which exhibit a higher spatial resolution than conventional cerebral electromagnetic measurements. To correctly estimate the cortical signals we used multicompartment head models (scalp, skull, dura mater, and cortex) constructed from individual MRI, a distributed source model, and a regularized linear inverse source estimates of cortical current density.

Chapters I–III discuss the simulation studies in which different main factors (signal-to-noise ratio, cortical activity duration, frequency band, etc.) are systematically imposed in the generation of test signals, and the errors in the estimated connectivity are evaluated by analysis of variance (ANOVA). In particular, we first explore the behavior of the most advanced estimators of effective and functional connectivity – SEM, DTF, dDTF, and PDC – in a simulation context and under different practical conditions.

For SEM, which involves the definition of an a priori connectivity model, the simulation study is designed to answer the following questions:

1. What is the influence of a variable signal-to-noise ratio (SNR) level on the accuracy of the pattern connectivity estimation obtained by SEM?

2. What is the amount of data necessary to get good accuracy of the estimation of connectivity between cortical areas?

3. How are the SEM performances degraded by an imprecise anatomical model formulation? Is it able to perform a good estimation of connectivity pattern when connections between the cortical areas are not correctly assumed? Which kind of errors should be avoided?

For the three multivariate estimators of functional connectivity – DTF, dDTF, and PDC – the experimental design focused on the following questions:

1. How are the connectivity pattern estimators influenced by different factors affecting the EEG recordings such as the signal-to-noise ratio and the amount of data available?

2. How do the estimators discriminate between the direct or indirect causality patterns?

3. What is the most effective method for estimating a connectivity model under the conditions usually encountered in standard EEG recordings?

These questions are addressed via simulations using predefined connectivity schemes linking several cortical areas. The estimation process retrieves the cortical connections between the areas under different experimental conditions. The connectivity patterns estimated by the four techniques are compared with those imposed on the simulated signals, and different error measures are computed and subjected to statistical multivariate analysis. The statistical analysis showed that during the simulations, SEM, DTF, and PDC estimators are able to estimate the imposed connectivity patterns under reasonable operating conditions. It was possible to conclude that the estimation of cortical connectivity can be performed not only with hemodynamic measurements, but also with EEG signals obtained from advanced computational techniques.

After giving a full description of the properties of these connectivity estimators for high resolution EEG recordings, the results of their application to human data relating to different experimental tasks such as finger tapping, Stroop test, and movement imagination are discussed. After the simulation tests, SEM, DTF, and PDC are applied to different sets of experimental data relating to motor and cognitive tasks (Chapters IV–VI). The motor task examined is a fast repetitive finger tapping, while the cognitive task involved recordings during the Stroop test, often employed in studies of selective attention, and found to be sensitive to prefrontal damage. The data employed are cortical estimates obtained from high-resolution EEG recordings using very advanced techniques which exhibit a higher spatial resolution than conventional cerebral electromagnetic measurements. We also briefly describe the high-resolution EEG techniques, including the use of a large number of scalp electrodes, which are realistic models of the head derived from structural magnetic resonance images (MRIs), and advanced processing methodologies related to solutions of linear inverse problems. The results of the estimation of the effective and functional connectivity from data recorded during finger tapping test, Stroop test and movement imagination test are presented.

One of the possible problems in the approach presented above is the hytpothesis that the gathered EEG data are stationary. However, this property of the data cannot be easily assumed. In fact, under this hypothesis, the methodology proposed for the DTF and PDC techniques is valid. Then the connectivity estimation methods for dealing with nonstationary EEG data are highly desirable. In Chapter VII we propose a methodology for the estimation of cortical connectivity extended to the time–frequency domain, based on the use of adaptive multivariate models. Such an approach allows extention of the connectivity analysis to nonstationary data and monitoring the rapid changes in the connectivity between cortical areas during an experimental task. The performances of the time-varying estimators are tested by means of simulations performed on the basis of a predefined connectivity scheme linking different cortical areas. Cortical connections between the areas are retrieved by the estimation process under different experimental conditions, and the results obtained for the different methods are evaluated by statistical analyses.

Finally, as an example of the results that can be obtained by this technique, the application of the simulation study to real data is proposed in Chapter VIII. For this purpose, we applied the time-varying technique to the cortical activity estimated in particular regions of interest (ROIs) of the cortex, and obtained high-resolution EEG recordings during the execution of a combined foot–lips movement in a group of normal subjects.

The experimental data presented here as practical results of the estimation of cortical connectivity in humans during motor and cognitive tasks were obtained from the Laboratory of High Resolution EEG of the University of Rome "La Sapienza" and from the Laboratory of Neuroelectric Imaging and Brain Computer Interface of the S. Lucia Foundation. In addition, part of the experimental data employed were provided by the Department of Biomedical Engineering, University of Minnesota, Minneapolis, USA, and by the Department of Psychology and Beckman Institute Biomedical Imaging Center, University of Illinois at Urbana-Champaign, Illinois, USA, in the framework of a scientific cooperation with the University of Rome.

CHAPTER 1

INTRODUCTION

The rapid growth of wireless local area and personal communication networks as well as sensor applications has led to a dramatic increase of interest in the regimes of radio frequency (RF)/microwave/millimeter-wave systems [1]. The 60 GHz band is of much interest as it is the band in which massive amounts of spectrum space (7 GHz) have been allocated worldwide for ultrafast wireless local communications, [2]. There are a number of multimedia applications for short-range communications, such as high-speed Internet access, video streaming, content downloads, and wireless data bus for cable replacement [3]. Such emerging applications with data rate greater than 2 Gb/s require real estate efficiency, low-cost manufacturing, and excellent performance achieved by a high-level of integration of embedded functions [2]. However, up to now the band around 60 GHz has been limited in use based on the difficulties associated with traditional radio propagation models and the cost of producing commercial products. Since the mid-1990's, many examples of monolithic-integrated circuit (MMIC) chipset have been reported for 60 GHz radio applications using gallium arsenide (GaAs), field effect transistor FET indium phosphide (InP), and pseudomorphic high electron mobility transistor (pHEMT) technologies [2]. Despite their commercial availability and their outstanding performance, these technologies struggle to enter the market because of their prohibitive cost and their limited capability to integrated advanced baseband processing. The combination of a low-cost, highly producible module technology, featuring low loss, and embedded miniaturized passive functions is required to enable a commercial use of the 60 GHz systems.

Considering the importance of the above-mentioned new applications at 60 GHz, a paradigm change in the system integration of millimeter-wave applications from high budgets and low volumes toward low costs and high-volumes is underway. With respect to millimeter-wave front-end modules, this leads to great challenges for both transceiver circuitry and antenna. Today's packaging technology of complex RF products, based on traditional chip and wire approach on single-layer ceramics and soft substrates, assembled into metallic boxes with complex geometries, hermetic sealing and ceramic coaxial feedthroughs cannot be the answers for high-volume production.

A complete reformation of this issue leads to the so-called three-dimensional (3D) SOP concept [1], wherein the traditional approach is replaced by a common package base architecture with a common multilayer substrate. This approach takes advantage of the passive integration, especially eliminating the individual passive components packages that usually occupy 90% of the system. Using SOP, the passive elements are converted to bare, thin-film components, only micrometers thick and embedded into the multilayer system package. Such thin-film components can be scaled anywhere from a thousandth to a millionth of their original packaged size [4]. Further, we can also eliminate the bulky integrated circuit (IC) packages by embedding their base chips in the SOP package [1]. In addition, the 3D integration, combining multiple layers of planar devices with a high-density of both in-plane and out-of-plane interconnects, is a promising approach to extending performance improvements beyond devices and interconnect scaling limits [5]. The 3D integration provides high device integration density, high interconnection count, compact mixed-signal interfaces, reduced-size wires and power consumption and novel miniaturized architectures. By layer stacking and adding another degree of freedom (vertical) to the functional connectivity, 3D integration can provide excellent functionality in a small footprint with reduced wiring lengths, thus reducing interconnect-driven signal delay and power consumption. Since the concept of 3D integration was first introduced in the 1980s [6–9], several 3D fabrication techniques (sequential [6–9], parallel [10], multilayer with buried structure (MLBS) [5]) have been demonstrated to realize 3D ICs. Also, 3D integration of microprocessors and memories (SRAM and DRAM) allows larger on-chip memory and on-chip interconnects, thus allowing for a substantial performance improvement in system due to improved memory hierarchy [11]. However, past 3D integration researches are still based on relatively low-density hybrid assembly technologies. The 3D integration using SOP technology allows the reduction of the overall substrate area required and a lower number of interfaces (lower loss). It also enables integration of matching structures, lumped and distributed passives, and antennas that is the so-called passive integration.

Both ceramic and organic technologies have been investigated in the 3D integration of miniaturized RF/microwave/millimeter-wave systems. Low temperature cofired ceramic technology (LTCC) has been widely used as a packaging material because of its process maturity/stability and its relatively high dielectric constant that enables a significant reduction in the module/function dimensions through the lamination of a large (up to 20–25) number of layers, making LTCC especially attractive for 3D integrated embedded components such as filters and antennas [12]. Although it has gained an increased popularity popularity in RF module implementation, LTCC suffers from a couple of main drawbacks such as the shrinkage of ceramic tapes during firing process and a coarse metal definition. More details will be discussed in Section 2.2. As an alternative, liquid crystal polymer (LCP) is an organic material that offers a unique combination of electrical, chemical, and mechanical properties, enabling low-cost high-frequency designs due to its ability to act as both the substrate and the package for flexible and conformal multilayer functions [4]. It is a fairly new, low-cost thermoplastic material, and its performance as an organic material is comparable to ceramic-based substrates that are widely used in RF and microwave applications (Table 1.1) [13].

LCP offers a large area processing capability that leads to tremendous cost reduction compared to LTCC substrates. Using vertical space allows the passive elements in RF front-ends to be efficiently integrated. However, processing challenges such as LCP-metal adhesion, bond registration and vias processing have delayed widespread LCP implementation in packaging.

TABLE 1.1: Material comparisons.

	ε_R	tan δ	CTE (ppm/°C)	COST
FR4	4	0.025	15–20	Very low
LTCC	3.9–9.1	0.0012–0.0063	3.4–7	Low/medium
LCP	2.9–3.2	0.002–0.0045	3–40	Low

With efficient electrical, mechanical and thermal performance in RF and microwave frequencies (sommtimes up to 110 GHz) has enabled the implementation of the 3D integration concept to the wireless multifunctional communication and sensor modules. Numerous publications [14–31] have dealt with the development of 3D LTCC passive components for microwave frequencies below the millimeter-wave range. However, recently reported structures occupy large areas without any miniaturization mechanism applied, while being hardly scalable to millimeter-wave dual-band, wideband, or multiplexing applications. In addition, these structures are designed below 50 GHz because short wavelengths and increased reflections at discontinuities require a technology with very tight tolerance.

On the basis of the given analysis, the three objectives underlying this book can be formulated as follows:

1. To develop one complete library of the 3D filters and duplexers in both microstrip and cavity configurations for compact, low-cost, and high-performance millimeter-wave front-end modules using LTCC technologies.

2. To demonstrate the feasibility of 60 GHz compact, high-data rate, high-gain, and directive antennas, that can be easily integrated in arrays configurations with 3D radiation enhancement topologies and integrated modules.

3. To provide a step-by-step description of the complete 3D integration of all passive building blocks, such as cavity duplexers and antennas, that enables the realization of "all-passive-integration" front-end solutions for compact 3D 60 GHz-band transceiver modules.

According to these objectives, the lecture is organized in a way that that the reader is presented with the general millimeter-wave module technologies in the first chapter along with their brief historical background and fundamental theory. A set of the considered technologies includes SOP, LTCC, T/R modules, and LCP packaging. The remaining chapters focus on the design of 60 GHz passive building blocks such as the filters and antennas and the complete 3-D integration of the developed passive building blocks.

CHAPTER 2

Background on Technologies for Millimeter-Wave Passive Front-Ends

2.1   3D INTEGRATED SOP CONCEPT

The system-on-package (SOP) is a multifunction package solution providing the necessary system functions that include analog, digital, radio frequency (RF), optical, and microelectronic-mechanical systems (MEMS). The SOP also allows for the efficient integration of complete passive RF front-end functional building blocks, such as filters and antennas. The recent development of thin-film RF materials makes it possible to bring the concept of SOP into the RF world and to meet stringent needs in wireless communication [4,32,33]. Major barriers that need to be addressed, especially for broadband applications, include wideband and low-loss interconnects; high-Q multilayer passives including resistors, inductors, and capacitors [34–36]; board-compatible embedded antennas and switches [36]; low-loss and low-cost boards; efficient partitioning of monolithic-integrated circuits (MMICs); low-crosstalk embedded transmission lines and single-mode packages as well as design rules for vertically integrated transceivers. In addition, there exists a gap in the area of hybrid computer-aided design (CAD) needed for novel functions that require fast and accurate modeling of electromagnetic, circuit, solid side, thermal, and mechanical effects. Multilayer ceramic [such as "Low-Temperature Cofired Ceramic" (LTCC)] and multilayer organic (MLO) structures [such as "Liquid Crystal Polymer"(LCP)] [35] are used to embed passives efficiently, including high-Q inductors, capacitors, matching networks, low-pass and band-pass filters, baluns, combiners, and antennas. The three dimensional (3D) design approach using multilayer topologies leads to high-quality and compact components that support multiband, wider bandwidth, and multistandard operation with high compactness and low cost.

Figure 2.1(a) illustrates the proposed 3D multilayer module concept [37]. Two stacked SOP multilayer substrates are used, and board-to-board, vertical transitions are ensured by use of micro ball grid array (μBGA) interconnects. Standard alignment equipment is used to stack the boards and thus provide a compact, high-performance, and low-cost assembly process. Multistepped cavities in the SOP boards enhance a tighter spacing for embedded RF active device (RF switch, RF receiver, and RF transmitter) chipsets and thus lead to significant volume reduction by minimizing the gap between the boards. Active devices can be flip-chipped and wirebonded. Cavities also provide an excellent

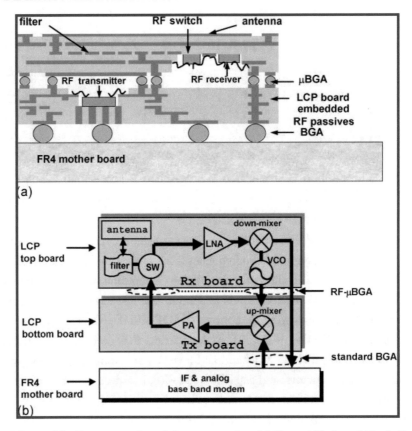

FIGURE 2.1: (a) 3D integrated module concept view (b) Rx and Tx board block diagram.

opportunity for the easy integration of RF MEMS devices, such as MEMS switches or tuners. Passive components, off-chip matching networks, embedded filters, and antennas can be easily implemented directly into the SOP boards by using multilayer technologies [38,39]. Standard BGA balls ensure the effective broadband interconnection of this high-density module with motherboards, such as the FR4 board. The top and bottom substrates are dedicated to the receiver and transmitter building blocks, respectively, of the RF front-end module.

Figure 2.1(b) shows the RF block diagram of each board. The receiver board includes an antenna, BPF, active switch, and an RF receiver chipset (low-noise amplifier (LNA), voltage control oscillator (VCO), and downconversion mixer). The transmitter board includes an RF transmitter chipset (up conversion mixer and power amplifier) and off-chip matching networks. Ground planes and vertical via walls are used to address the isolation issues between the transmitter and the receiver functional blocks. Arrays of vertical vias are added into the transmitter board to achieve better thermal management.

2.2   LTCC MULTILAYER TECHNOLOGY

LTCC stands for a ceramic substrate system, which is applied in electronic circuits as a cost-effective and competitive substrate technology with a nearly arbitrary number of layers. In general, printed gold and silver conductors or alloys with platinum or palladium are used. However copper conductors are also available. The metallization pastes will be screen-printed layer-by-layer upon the unfired and "green" ceramic foil, followed by stacking and laminating under pressure. The multilayer ceramic stack will be fired (sintered) in the final manufacturing step. The sintering temperature is below 900°C for the LTCC glass-ceramic. This relatively low temperature enables the cofiring of gold and silver conductors. The melting of Au and Ag are at 960°C and 1100°C, respectively.

The LTCC slurry is a mixture of recrystallized glass and ceramic powder in binders and organic solvents. It is cast under "doctor blades" to obtain a certain tape thickness. The dried tape is coiled on a carrier tape and ready for production. In contrast, High temperature cofired ceramic (HTCC) is an aluminum oxide substrate, also called alumina or Al_2O_3. The sintering temperature of aluminum is 1600°C, which only allows cofiring conductors with a higher melting point such as tungsten (3370°C) and molybdenum (2623°C). The drawback of these conductors is their lower conductivities, which results in higher waveguide losses. The low line losses as well as competitive manufacturing costs are an advantage of LTCC for RF and microwave applications.

However, one of the main drawbacks of LTCC limiting its accuracy is the shrinkage of ceramic tapes during the firing process. Typically, the tapes shrink between 12–16% in the horizontal dimensions and 15–25% in the vertical. Typical shrinkage tolerances are ±0.2% and ±0.5% for both directions, respectively, and are dependent on the amount of conductor material on every layer. To eliminate shrinkage altogether, some manufacturers promoted tape on substrate technology (TOS) [40,41]. Shrinkage is virtually eliminated by laminating and firing each layer of tape on a substrate made of Al_2O_3, BeO or AiN. While this eliminates any component assembly alignment problems associated with the shrinkage, TOS is a serial process requiring expensive ceramic substrate carriers, and thus resulting in higher manufacturing costs. Another approach to solve the shrinkage problem was the development of low-temperature cofired ceramics on metal (LTCC-M) [42] with a specially formulated multilayer ceramic structure attached to a metal core. In this case, the ceramic firing and core attachment process occur in one and the same step, which is a more cost-effective solution than TOS. The resulting structure exhibits virtually no shrinkage in the plane of the substrate. Vertical shrinkage is, however, still an issue. Another basic limitation of the LTCC process is a coarse metal definition. The minimal line width and spaces that may be achieved with normal thick film techniques are typically 125 μm with a typical tolerance of ±25 μm. The minimal conductor thickness is around 25 μm.

As long as the frequency is low or the system specifications are relaxed, LTCC may be successfully used. At the V-band, however, the application of this technology for practical integration of high-performance passive structures using planar transmission line media [microstrip and coplanar waveguide (CPW)] seems to be difficult. For microstrip realizations, the dielectric tapes have to feature a very consistent and predictable thickness. The thin dielectric layers necessary for operation at the V-band impose an accuracy limit on the thickness of LTCC tapes that is difficult to achieve. Furthermore, the use of thin dielectric tapes involves the need for precisely defined narrow strips and spaces required for the repeatable performance of many passive structures. Taking into account the high dielectric constant of LTCC tapes, the realization of precise CPW structures with appropriately narrow ground–ground spacing, necessary for avoiding substantial coupling and leakage to substrate modes at the V-band, also seems to be very challenging. In the 3D module integration, the coefficient of thermal expansion (CTE) is also an important parameter as it affects the Si-based ICs that are integrated. The substrate is expected to exhibit CTE values close to Si ($\sim$4.2 ppm/$^\circ$C) in order to avoid deformations such as cracks, delaminations, etc. between the substrate and the attached components due to shrinkage mismatch. LTCC technology, which treats substrate thick-film cofiring and device integration optimally, covers most of these requirements and thus is one of the few material systems employed for fabrication of 3D module systems.

To overcome the limitations of previous LTCC processes at millimeter-wave frequencies, an LTCC 044 SiO_2-B_2O_3 glass was recently developed and proposed. The relative permittivity (ε_r) of the substrate is 5.4 and its loss tangent (tan δ) is 0.0015 at 35 GHz. The dielectric layer thickness per layer is 100 μm, and the metal thickness is 9 μm. The resistivity of metal (silver trace) is determined to be 2.7×10^{-8} Ωm. The tolerance of XY shrinkage has been investigated to be $\pm$5%. To eliminate the shrinkage effect, a novel composite LTCC of a high dielectric constant ($\varepsilon_r \sim 7.3$) and low dielectric constant ($\varepsilon_r \sim 7.1$) has been proposed. The very mature multilayer fabrication capabilities of this novel process (ε_r = 7.1 and 7.3, tan δ = 0.0019 and 0.0024, metal layer thickness: 9 μm, dielectric layer thickness: 53 μm, minimum metal line width and spacing: up to 75 μm) make it one of the leading competitive solutions to meet millimeter-wave design requirements in terms of the physical dimensions of the integrated passives. The proposed LTCC technologies can be very promising for the implementation of integrated cavities or waveguides at V-band or above with accurate via processing, substantially relaxing requirements on metal pattern accuracy.

2.3   60 GHz TRANSMITTER/RECEIVER MODULES

With the availability of 7 GHz of unlicensed spectra around 60 GHz, there is a growing interest in using this resource for new consumer applications, such as high-speed Internet access, streaming content downloads, and wireless data bust for cable replacement, requiring very high-data-rate wireless transmission. The targeted data rate for these applications is greater than 2 Gb/s, potentially reaching 20–40 Gb/sec. Although the excessively high path loss at 60 GHz resulting from oxygen absorption precludes communication over distances greater than a few tens of meters, short-range wireless personal area networks (WPAN) actually benefit from the attenuation, which provides extra spatial isolation and higher implicit security. Furthermore, because of oxygen absorption, Federal

Communications Commission (FCC) regulations allow for up to 40 dBm equivalent isotropically radiated power (EIRP) for transmission, which is significantly higher than what is available for other wireless local area network (WLAN)/WPAN standards. The wide bandwidth and high allowable transmitting power at 60 GHz enable multigigabit-per-second wireless transmission over typical indoor distances (∼10 m). Moving to higher frequencies also reduces the form factor of the antennas, as antenna dimensions are inversely proportional to carrier frequency. Therefore, for a fixed area, more antennas can be used, and the antenna array can increase the antenna gain and help direct the electromagnetic energy to the intended target.

Assuming simple line-of-sight free-space communication, the Friis propagation is given by

$$\frac{P_r}{P_t} = \frac{D_1 D_2 \lambda^2}{(4\pi R)^2} \tag{2.1}$$

where the received power, P_r, normalized to the transmitted power, P_t, is seen to depend on the transmitter and receiver antenna directivities, $D_{1,2}$; the distance between the transmitter and receiver, R, and wavelength, λ. Assuming a single omnidirectional antenna at the transmitter and receiver, the received power, P_r at a distance, R, decreases with increasing frequency, and an additional 20 dB loss is expected for a system operating at 60 GHz, compared to 6 GHz. This additional loss would render the system incapable of delivering gigabit-per-second data rates at 10 m given the limited output power and high noise figure of a 60 GHz transceiver implementation.

Fortunately, the antenna directivities, $D_{1,2}$, can be improved. While it is impossible to increase the antenna gain for a single antenna, it is more desirable to increase the directivity by employing an antenna array. For a fixed antenna aperture size, A, the directivity is given by

$$D = \frac{4\pi A}{\lambda^2} \tag{2.2}$$

To implement a 60 GHz front-end module, several technologies have been investigated. In particular, GaAs FET technology has evolved to the point where 60 GHz GaAs MMICs are ready for production [43]. GaAs-based 60 GHz devices such as low-noise amplifiers, high-power amplifiers, multipliers, and switches can nowadays be ordered in large quantities in die form at prices $10–20 per piece [44]. For application in WLAN equipment, however, this might still be too expensive. An alternative technology based on silicon germanium (SiGe) promises to provide relatively low-cost millimeter wave front-end MMICs while simultaneously maintaining the favorable performance of GaAs. Transceiver circuits using SiGe have been demonstrated to operate at 60 GHz with good performance [45]. However, digital complementary metal-oxide semiconductor (CMOS) technology is the lowest cost option, and with its rapid improvement due to continual scaling, CMOS technology is becoming a viable option to address the millimeter-wave market. Today, 90 nm bulk CMOS technology is used to implement power and low-noise amplifiers at 60 GHz [46]. Future bulk CMOS process at the 65 nm node are expected to provide even higher gain at lower power consumption. A 60 GHz receiver front-end realized in 130 nm CMOS technology has been success-fully implemented with low power consumption and compact size [47]. Also, the development of 60 GHz-band T/R modules for stringent system specifications has been demonstrated in a system-in-package (SIP) transmitter integrating LTCC patch-arrayed antennas [48] and compact wireless transceiver modules for gigabit data-rate transmission [49–51]. However, the previously reported transmitter and receiver modules could suffer from the spurious signal and image signal, because only antennas are integrated into modules without using any band-select filters or duplexers in passive front-ends. Moreover, two separate antennas for the Tx and Rx channels are used and occupy a large area, which contradicts the size requirements of compact 60 GHz modules.

Among some of the major design concerns for the design and development of 3D integrated modules in mm-frequencies are the front-end power consumption levels, that locally lead to very high temperatures with detrimental impact on module's operation [52]. Thus, it would be necessary to optimize the thermal performance of the front-end module and to improve its reliability at a reduced overall cost. The thermal performance of front-end modules incorporating LTCC substrate has been investigated at L band [52]. It shows that the peak junction temperature for the power amplifier with LTCC substrate and silver paste metallization is around 130.1°C, ∼51% higher compared to the baseline case with two-layer organic substrate. By increasing the metal thermal conductivity from 90 (silver paste) to 150, 250, and 350 W/mK, a significant drop in peak temperature occurs, indicating its impact on PA's overall thermal performance. The thickness of the top metal (10 μm versus 30 μm) contributes only to 5–8% change in peak junction temperature. In addition, it has been shown that the vias placed under the die are the dominating mechanism removing the heat from the die through the substrate, while the outer vias have no impact on the PA overall thermal performance.

CHAPTER 3

Three-Dimensional Packaging in Multilayer Organic Substrates

3.1   MULTILAYER LCP SUBSTRATES

As the operating frequencies of radio frequency (RF) systems continue to rise, system reliability becomes increasingly dependent on hermetic or near hermetic packaging materials. Higher frequencies lead to smaller circuits, and low material expansion (which is related to water absorption) becomes more important for circuit reliability and for maintaining stable dielectric properties. Equally important is the ability to integrate these materials easily and inexpensively with different system components. The best packaging materials in terms of hermeticity are metals, ceramics, and glass. However, these materials often give way to cheaper polymer packages such as injection-molded plastics or glob top epoxies when the cost is a driving factor. Plastic packages are suitable when cost and ease of fabrication are considered, but they are not very good at keeping out water and water vapor. Ideally, a hermetic polymer should have inexpensive material and fabrication cost and still function as a good microwave and millimeter-wave package.

An organic material called liquid crystal polymer (LCP) nearly satisfies these criteria. The dielectric properties of LCP and the performance of various transmission lines on LCP substrates were recently characterized up to 110 GHz [53] through the use of ring and cavity resonators, that derived the following characteristics: $\varepsilon_r = 3.16 \pm 0.05$ from 31.53 to 104.06 GHz and $\tan \delta < 0.0049$ up to 97 GHz. The low water absorption of LCP makes it stable across a wide range of environments by preventing changes in the relative dielectric constant (ε_r) and loss tangent ($\tan \delta$) [D. C. Thompson, M. M. Tentzeris and J. Papapolymerou, "Experimental Analysis of the Water Absorption Effects on RF/mm-wave Active/Passive Circuits Packaged in Multilayer Organic Substrates", IEEE Transactions on Advanced Packaging, Vol. 30, No. 3, pp.551—557, August 2007]. The LCP material processing is still in infancy, but because of LCP's capability to do reel-to-reel processing, it is expected that production costs will continue to decrease. At the same time, the flexibility and relatively low processing temperatures of the material ($\sim 280^\circ$C—290°C) enables applications such as conformal antennas and integration of microelectromechanical system (MEMS) devices such as low-loss RF switches.

In addition, LCP-based multilayer modules are easy to fabricate due to the existing two types of LCP material with similar electrical characteristics and different melting temperatures. High

TABLE 3.1: Comparison of transverse coefficient of thermal expansion.

	CTE (ppm/°C)
LCP	3–40
Cu	16.8
Au	14.3
Si	4.2
GaAs	5.8
SiGe	3.4–5

melting temperature LCP (xxxx) can be used as a core layer, while low melting temperature LCP (290°C) can be used as a bond ply. Thus, vertically integrated designs may be realized similar to those in low temperature cofired ceramic technology (LTCC). An additional benefit in multilayer LCP front-ends is the enhanced functionality provided by the low dielectric constant, especially for planar antennas printed on the top layer of an all-LCP modules.

Use of LCP as a microwave circuit substrate is not a new idea. It has been around in thin-film form since the early 1990s when it was first recognized as a candidate for microwave applications [54–56]. However, early LCP films would easily tear and were difficult to process. Film uniformity was not acceptable and poor LCP to metal adhesion and failure to produce reliable plated through holes (PTHs) in LCP limited the capabilities for manufacturing circuits on it. Devising and optimizing LCP surface treatments and via drilling and de-smearing techniques were also necessary in order to bring the material to a state where circuits on it could be manufactured with confidence. A biaxial die extrusion process was developed [56–58], which solved the tearing problems by giving the material uniform strength and it also created additional processing benefits. It was discovered that by controlling the angle and rate of LCP extrusion though the biaxial die, the x–y coefficient of thermal expansion (CTE) could be controlled approximately between 0 and 40 ppm/°C. Thus, this unique process can achieve a thermal expansion match in x–y plane with many commonly used materials. Table 3.1 shows how the transverse CTE of LCP can be engineered to match both metals and semiconductors used in high-frequency systems.

LCP's z-axis CTE is considerably higher ($\sim$105 ppm/°C), but due to the thin layer of LCP used, the absolute z-dimension difference between LCP and a 2-mil high copper PTH is less than 0.5 μm within a $\pm$100°C temperature range. This makes the z-axis expansion a minimal concern until very thick multilayer modules come into consideration.

3.2   RF MEMS PACKAGING USING MULTILAYER LCP SUBSTRATES

In this section, the concept of packaging numerous devices with a standard thin-film LCP layer is presented. A 4-mil nonmetallized LCP superstrate layer with depth-controlled laser micromachined cavities is investigated as a package. This technique is demonstrated by creating packages for air-bridge RF MEMS switches [G. Wang, D. Thompson, M. M. Tentzeris and J. Papapolymerou, "Low Cost RF MEMS Switches Using LCP Substrate", Procs. of the 2004 European Microwave Symposium, pp. 1441—1444, Amsterdam, The Netherlands, October 2004.]. The switch membranes are only about 3 μm above the base substrate that allows a cavity with plenty of clearance to be laser drilled in the LCP superstrate layer. A cavity depth of 2 mils ($\sim$51 μm, half of the superstrate thickness) was chosen for the MEMS package cavities.

3.2.1   Package Fabrication

A CO_2 engraving laser with a 10 μm wavelength was used to drill holes in the LCP superstrate layer. The CO_2 laser was selected for this job due to its high power and the corresponding fast cutting rate. Circles were cut out in the four corners for pin alignment and square or rectangular windows were removed in specified locations for the probe feedthroughs. The alignment holes and feedthrough holes were drawn in AUTOCAD, programmed into the laser software, and the cuts were made concurrently in a single laser run.

An excimer laser was used to micromachined depth-controlled cavities in the desired locations. The stage was aligned to the already cut holes from the CO_2 laser and the laser was again programmed to fire in a predetermined pattern. The optical alignment was limited by the large aperture size, but the accuracy was estimated to be within 100 μm at the worst case. The laser cavity dimensions were large enough so that the potential alignment error was of no concern. With smaller apertures, an alignment better than 10 μm can be accomplished with the excimer laser.

The laser power and the number of pulses were tuned to provide the desired ablation depth into the LCP superstrate. We arbitrarily chose to make cavity depths half of the substrate thickness (2 mil deep cavities). Shallower or deeper cavities are possible by varying the laser power and the number of pulses. A custom brass aperture with a rectangular hole was used to shape the beam to the desired shape and size of the cavity. The aperture size of 12 mm × 5 mm was demagnified five times to create a cavity 2.4 mm wide × 1 mm long. After machining the cavities, the depth was checked with a microscope connected to a digital z-axis focus readout with accuracy to the nearest tenth of a micrometer. The depth across the bottom of the cavities was not completely uniform due to some small burn marks on the laser optics, but it was within ±5 μm of the desired depth across the entire cavity.

The completed package layers were made such that the alignment holes corresponded to the same location as those on the through-reflect-line (TRL) calibration lines and also on the MEMS

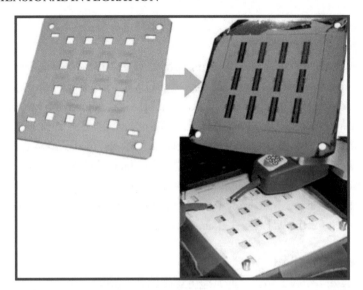

FIGURE 3.1: Stack package and MEMS substrates over alignment pins. (*Top left*) LCP superstrate packaging layer with holes for alignment and probe feedthroughs. The packaged cavities between each set of probing holes are visible due to LCP becoming partially transparent at a 2 mil thickness. (*Top right*) Conductor-backed finite ground coplanar (CB-FGC) transmission lines with air-bridge RF MEMS switches in the center of the transmission lines. (*Bottom right*) Both layers stacked on alignment fixture and probed through the feedthrough windows.

switch samples. The package was aligned and stacked over the MEMS substrate with the assistance of four alignment pins as seen in Fig. 3.1.

3.2.2   RF MEMS Switch Performance with Packaged Cavities

The MEMS switches were comprised of a 2 μm thick electroplated gold doubly supported air-bridge layer suspended approximately 3 μm above the lower metal layer. The 100 μm × 200 μm membrane were suspended over the signal line of a conductor-backed finite ground coplanar (CB-FGC) transmission line and anchored to the ground planes on both sides. In the default state, the membrane is up, in which case full signal transmission should take place. When a DC actuation voltage is introduced, the membrane is flexed down into contact with a thin silicon nitride layer between the two metal layers and creates a capacitive short circuit that blocks signal transmission. A picture of a fabricated MEMS switch is shown in Fig. 3.2.

As MEMS switches are by nature fragile, an iterative measurement procedure was undertaken. First, the switches were measured in air to provide a base measurement case. The second and third measurements were done with the package layer aligned and held in contact with the base substrate. The first packaging iteration was done by gently holding the package layer down over the MEMS substrate with tape. When the switches continued to operate with the package layer in place, this

FIGURE 3.2: Fabricated RF MEMS switch.

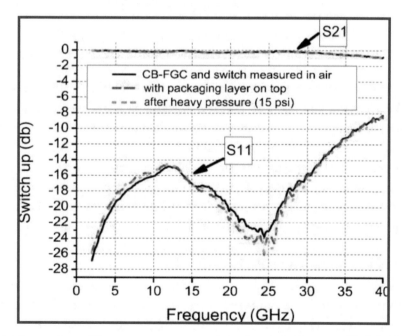

FIGURE 3.3: Comparison of S-parameter measurements of an air-bridge type CB-FGC MEMS switch in the "UP" state. (*Case 1*) The switch is measured in open air. (*Case 2*) The packaging layer is brought down and taped into hard contact and measured. (*Case 3*) A top metal press plate and a 15 lbwt are put on top of the packaging layer (15 psi) to simulate bonding pressure. The weight and press plate are then removed and the switch is remeasured.

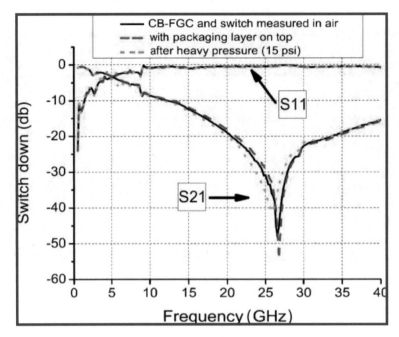

FIGURE 3.4: Comparison of S-parameter measurements of an air-bridge type CB-FGC MEMS switch in the "DOWN" stage. The three measurement cases shown are identical to those shown in Fig. 3.3.

ensured that the alignment of the package cavities was successful. Finally, the top metal plate was placed over the alignment pins and a 15 lbwt was balanced on top of the samples to simulate the pressure from a bonding process. The plate was removed and the samples were remeasured. Results for these measurements are shown in Figs. 3.3 and 3.4.

The S-parameters of the packaged switch and the nonpackaged switch are nearly identical in both the up and down states. For example, the variation between the three measurement cases for S21 in the "UP" state only varies by an average of 0.032 dB across the entire measurement band. The other S-parameter comparisons with and without the package layer are very similar.

3.2.3   Transmission Lines with Package Cavities

To show the effects of the packaging layer and cavity on a simple transmission line, the switch membrane was physically removed and the circuit remeasured. The results of the bare transmission line with and without the packaging layer are shown in Fig. 3.5. As expected from these simulations, the cases with and without the packaging layer are very similar.

3.3   ACTIVE DEVICE PACKAGING USING MULTILAYER LCP SUBSTRATES

Active devices, specifically GaAs MMICs, are robust to humidity and temperature testing. The gold metallization on GaAs chips relieves several of the problems that plague Si MMICs, which have

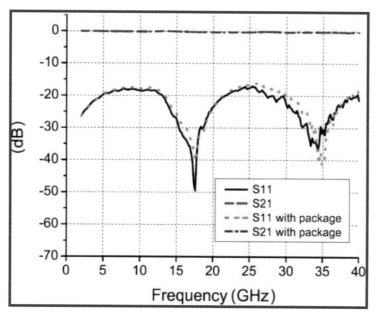

FIGURE 3.5: Comparison of S-parameter measurements of the MEMS switch transmission line after the switch was physically removed. The cases with the package and without the package layer are nearly the same.

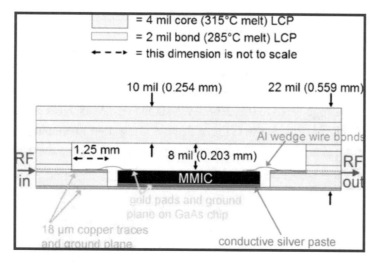

FIGURE 3.6: Pictorial side view of the package stackup.

aluminum contacts. However, for a reliable and long-term operation, a substantial sealed package is still desired to protect GaAs MMICs from the environment. In addition, to create compact, inexpensive RF modules, new packaging concepts and convenient integration techniques of combining passive and active devices are required. One such technique, which operates similar to the low-temperature co-fired ceramic (LTCC) fabrication flow, but whose laminated temperature is low enough for embedding chips, is bonding multiple thin-film LCP substrates into a package with embedded cavities for MEMS or monolithic microwave integrated circuits (MMICs) [59].

3.3.1   Embedded MMIC Concept

The idea for embedding a MMIC in a multilayer dielectric substrate/package for creating compact RF modules is not new. LTCC is a material technology that allows the space savings of embedding passive elements on many vertically connected layers. Unfortunately, LTCC has a firing temperature of 850 °C, which means that the inclusion of MMICs must be done with some external assembly process after firing. This can involve soldering plastic leaded chips onto the top layer or using other methods to embed chips in cavities between already fired LTCC boards. Since LCP has a lamination temperature of 285 °C, chips can be included directly inside the LCP layer stackup and laminated/packaged during the same thermocompression bonding process that seals the rest of the passive element layers together. Two issues that are important for the reliability of active devices are coefficient of thermal expansion (CTE) matching at the semiconductor connection points, and thermal heat dissipation. To prove the concept of a robust multilayer LCP packaged MMIC and to address these issues, the package design shown in Fig. 3.6 was devised.

The coefficient of thermal expansion (CTE) of the chip's gold ground plane [14.4 (ppm/°C)] is well matched to the special inorganic silver epoxy adhesive and copper layers [both with 17 (ppm/°C)] to which its base is attached. In addition, this contact location is excellent for heat dissipation as the chip is directly connected to the large copper RF ground plane. However, to be realistic about the CTE match, the base of the chip may not be the most sensitive area of concern. It is more likely to be of importance in locations where the chip contacts connect to feed lines. Fortunately, LCP's CTE in the x–y plane can be engineered to match both metals and semiconductors at the expense of slight changes to the z-CTE. LCP with the CTE of 5 ppm/°C are used for semiconductor attachment and layers with the CTE of 17 ppm/°C are used in layers where matching the copper metallization. For reference, copper has a CTE of 17 ppm/°C and GaAs has a CTE of 5.8 ppm/°C.

3.3.2   MMIC Package Fabrication

Several laser micromachining process steps were used to create the multilayer LCP package. First, an excimer laser was used to form the chip cavity in the base substrate layer by ablating LCP down to the 18 μm copper ground plane. The standard 4 mil GaAs MMIC thickness is the same as an off-the-shelf LCP thickness so that the top of the chip is coplanar with feeding transmission lines on the LCP substrate. A Hittite HMC342 13–25 GHz low noise amplifier and an off-chip parallel plate bypass capacitor from Presidio Components Inc. were then affixed to the ground plane with an inorganic high temperature silver paste. These assembly steps are shown graphically in Fig. 3.7.

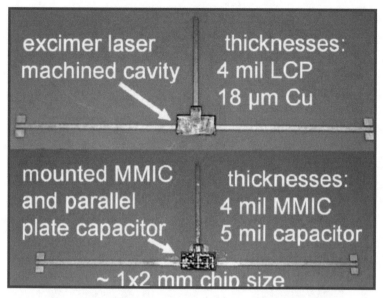

FIGURE 3.7: Comparison of the LCP laser machined base layer before and after the MMIC and parallel plate capacitor were mounted with an inorganic silver paste and wire bonded to the feed lines.

The superstrate packaging layers were machined with a CO_2 laser to form square holes in some layers for the chip cavity while leaving other layers solid to create a sealed cavity after lamination. All the layers including the base substrate had laser cut alignment holes in the same relative locations to enable precise stacking on an aluminum bonding fixture. The final laminated package on the fixture with the top press plate removed is shown Fig. 3.8.

3.3.3   MMIC Package Testing

The important proof-of-concept for the packaging of the MMIC is that a seal can be created around the 18-μm thick feeding transmission lines. These transmission lines pass directly through the side of the package stackup and require a 2 mil (50 μm) low melting temperature LCP bond layer to melt and conform around them to create a seal. Figure 3.9 shows a scale representation of the height of LCP's default metallization to the height of the bond ply. A closeup picture of the actual transmission line feedthrough, which demonstrates the ability of the LCP material to conform around the transmission line, is shown in Fig. 3.9.

To test the package seal, the packaged MMIC was submersed in water for 48 h. The sample was held on edge while underwater to encourage any potential cavity leaks to be breached. A through-reflect-line (TRL) calibration was performed with an identical alternate sample so that the measurement of the packaged chip could be made immediately upon removal from the water. The gain of the packaged MMIC was then measured and compared with the gain before the submersion test. The results of this test are shown in Fig. 3.10.

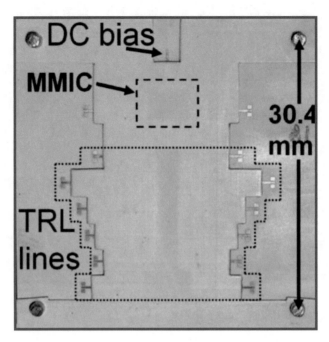

FIGURE 3.8: Top view of the 13–25 GHz GaAs MMIC packaged in multiple thin layers of LCP.

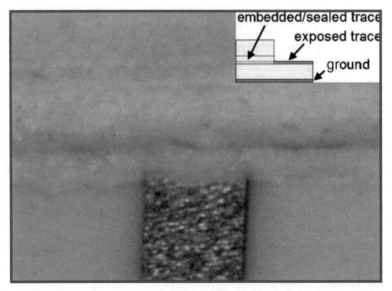

FIGURE 3.9: LCP transmission line (18 μm thick) passing directly through the side of a bonded superstrate package stackup.

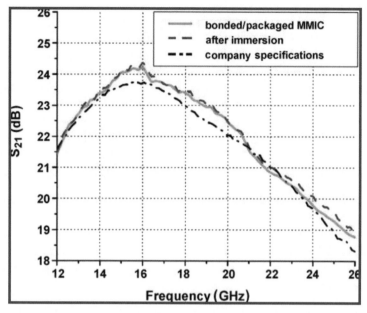

FIGURE 3.10: Gain measurement of the Hittite HMC342 13–25 GHz LNA. The first measurement was of the packaged/bonded MMIC. The second measurement was done immediately after the packaged MMIC was submerged in water on edge for 48 h. The match of the measurements demonstrates as successful seal by the LCP package. The minimal water absorption into the package shows no significant effect on the MMIC performance.

The gain measurement in the before/after states is identical, indicating that the multilayer LCP MMIC package method can be used successfully for packaging active devices.

3.4   THREE-DIMENSIONAL PAPER-BASED MODULES FOR RFID/SENSING APPLICATIONS

As the demand for low-cost, flexible and efficient electronics increases, the materials and integration technologies become more critical and face many challenges, especially with the ever growing interest for "cognitive intelligence" and wireless applications, such as radio frequency identification (RFID) and wireless local area networks (WLAN). Paper has been considered as one of the best organic-substrate candidates for ultrahigh frequency (UHF) and microwave applications such as RFID/sensing. It is not only environmentally friendly, but can also undergo large reel-to-reel processing. In terms of mass production and increased demand, this makes paper the lowest cost material made. Paper also has low surface profile with appropriate coating. This is very crucial since fast printing processes, such as direct write methodologies, can be utilized instead of metal etching techniques. A fast process, like inkjet printing, can be used efficiently to print electronics on/in paper substrates.

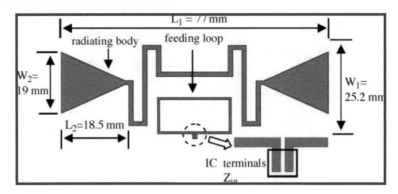

FIGURE 3.11: Inductively coupled feeding RFID tag module configuration.

First of all, the RF characteristics of the paper-based substrate have been recently studied by using the cavity resonator method and the transmission line method to characterize the dielectric constant (ε_r) and loss tangent (tan δ) of the substrate [60]. The results show $\varepsilon_r = 1.6$ at Ka band and tan $\delta < 0.082$ up to 2.4 GHz. Then, a UHF RFID tag module was developed with the inkjet-printing technology that could function as a technology for much simpler and faster fabrication on/in paper. Most available commercial RFID tags are passive, and the antenna translates electromagnetic waves from the reader into power supplied to the IC. Thus, a conjugate impedance matching between antenna and the tag IC is highly essential to power up the IC and maximize the effective range, and so an inductively coupled feeding structure is an effective way for impedance matching.

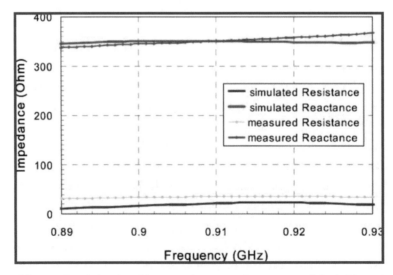

FIGURE 3.12: Measured and simulated input resistance and reactance of the inkjet-printed RFID tag.

As a benchmark of this approach, an inductively coupled feeding RFID tag module was inkjet-printed and its configuration is shown in Fig. 3.11. The target RFID IC was EPC Gen2 RFID ASIC IC, which has a stable impedance performance of 16-j350 Ohm over $902 - 928\,\text{MHz}$, covering the North America UHF RFID Band. The substrate was the previously characterized paper. To ensure that the ink droplets overlap sufficiently, a 25 μm drop spacing was selected. After inkjet printing, a low-temperature sintering step guaranteed a continuous metal conductor, providing a good percolation channel for the conduction electrons to flow. The measured and simulated RFID tag input impedance are shown in Fig. 3.12. Since the conductivity of the conductive ink varies from $0.4\text{--}2.5 \times 10^7$ siemens/m depending on the curing temperature [57], the input resistance was slightly higher due to the additional metal loss introduced. Overall, a very good agreement was observed over the frequency band of interest. It is quite clear that paper will become a commonly used material in 3D integration and packaging of the future, due to its ultra-low cost, ease of multilayer lamination in very low temperatures ($\sim 150°\text{C}$) and it enevironmentally-friendly ("gree") characteristics, especially in the UHF and Wireless frequency bands.

CHAPTER 4

Microstrip-Type Integrated Filters

4.1 PATCH RESONATOR FILTERS AND DUPLEXERS

4.1.1 Single Patch Resonator

Integrating filter-on-package in low-temperature cofired ceramic (LTCC) multilayer technology is a very attractive option for radio frequency (RF) front-ends up to the millimeter-wave frequency range in terms of both miniaturization by vertical deployment of filter elements and reduction of the number of components and assembly cost by eliminating the demand for discrete filters. In millimeter-wave frequencies, the bandpass filters are commonly realized using slotted patch resonators because of their miniaturized size and their excellent compromise among size, power handling, and easy-to-design layout [17]. In this section, the design of a single-pole slotted patch filter is presented for two operating frequency bands (38–40 GHz and 58–60 GHz). This design can be easily generalized to multiband applications, especially for portable wireless modules that the size and weight is of paramount importance. Its major advantages are its capability of high-Q structures in vertical stackups and the easy addition of multiple stages for high-selectivity applications.

Figure 4.1(a) and (b) shows a top-view comparison between a basic half-wavelength ($\lambda/2$) square patch resonator ($L \times L = 0.996\,mm \times 0.996\,mm$) [61] and the new configuration ($L \times L = 0.616\,mm \times 0.616\,mm$), respectively, that is capable of providing good tradeoffs between miniaturization and power handling. Side views and the photographs of the 60 GHz resonators are shown in Figs. 4.2 and 4.3, respectively. In the conventional design of a $\lambda/2$ square patch, the planar single-mode patch and microstrip feedlines are located on metal 3 (M3 in Fig. 4.2) and they use the end-gap capacitive coupling between the feedlines and the resonator itself to achieve 3% 3-dB bandwidth and <3 dB insertion loss around the center frequency of 60 GHz. However, the required coupling capacitances to obtain design specifications could not be achieved because of the LTCC design rule limitations.

To maximize the coupling strength while minimizing the effects of the fabrication, the proposed novel structure takes advantage of the vertical deployment of filter elements by placing the feedlines and the resonator into different vertical metal layers, as shown in Fig. 4.2. This transition also introduces a 7.6% frequency downshift resulting from the additional capacitive coupling effect compared to the basic $\lambda/2$ square patch resonator [Fig. 4.1(a)] directly attached by feedlines. Transverse cuts have been added on each side of the patch to achieve significant miniaturization of the patch by contributing an additional inductance. Figure 4.4 shows the simulated response for the

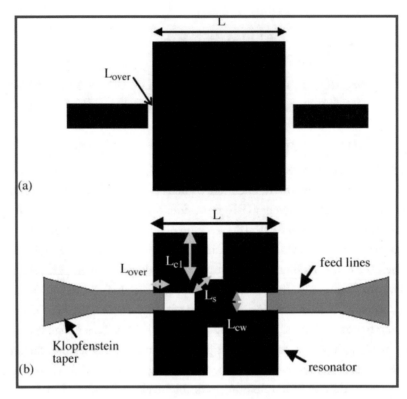

FIGURE 4.1: Top view of (a) conventional λ/2 square patch (b) Miniaturized patch resonator.

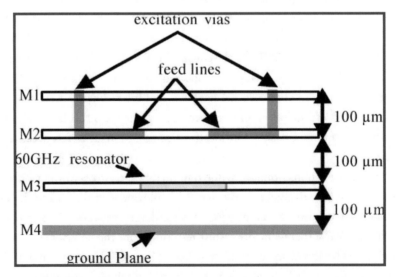

FIGURE 4.2: Side view of 60 GHz slotted 3D patch resonator.

FIGURE 4.3: Photograph of the fabricated filters with coplanar waveguide (CPW) pads at 60 GHz.

center frequency and the insertion loss as the length of cuts [L_{CL} in Fig. 4.1(b)] increases, while the fixed width of cuts [$L_{CW} = L/8$ in Fig. 4.1(b)] is determined by the fabrication tolerance. It can be observed that the operating frequency range shifts further downward about 33% as the length of the cut [L_{CL} in Fig. 4.1(b)] increases by approximately 379 µm. Additional miniaturization is limited by the minimum distance [L_S in Fig. 4.1(b)] between the corners of adjacent orthogonal cuts. Meanwhile, as the operating frequency decreases, the shunt conductance in the equivalent circuit of the single patch also decreases because its value is reciprocal to the exponential function of the operating frequency [62]. This fact additionally causes the reduction of radiation loss since it is proportionally related to the conductance in the absence of conductor loss [62]. Therefore, insertion loss at resonance is improved from 2.27 dB to 1.06 dB by an increase of L_{CL} in Fig. 4.1(b).

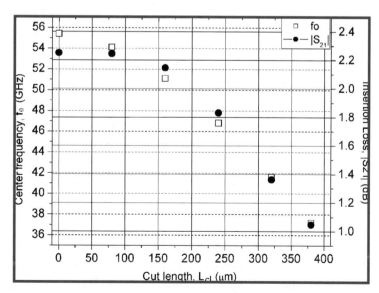

FIGURE 4.4: Simulated responses of center frequency (f_0) and insertion loss ($|S21|$) as a function of transverse cut (L_{CL}).

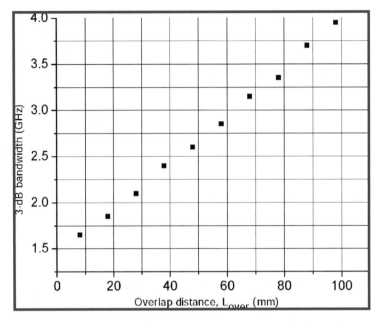

FIGURE 4.5: Simulated 3-dB bandwidth as function of overlap distance of 60 GHz slotted patch resonator.

The patch size is reduced significantly from 0.996 to 0.616 mm. The modification of bandwidth resulting from the patch's miniaturization can be compensated by adjusting the overlap distance (L_{over}). Figure 4.5 shows the simulated response for the 3-dB bandwidth as L_{over} increases. It is observed that the 3-dB bandwidth increases almost linearly as L_{over} increases because of a stronger coupling effect; L_{over} is determined to be 18 μm corresponding to the 1.85 GHz 3-dB bandwidth.

The proposed embedded microstrip line filters can be easily excited through vias connecting the coplanar waveguide (CPW) signal pads on the top metal layer (M1 in Fig. 4.2), reducing the paraisitc radiation loss compared to conventional microstrip lines on the top (surface) layer. As shown in Fig. 4.1(b), Klopfenstein impedance tapers are used to connect the 50 Ω feeding line and the via pad on metal 2 (M2 in Fig. 4.2). The overlap ($L_{\text{over}} \approx L/31$) and transverse cuts ($L_{\text{CW}} \approx L/8$, $L_{\text{CL}} \approx L/3.26$) have been finally determined to achieve desired filter characteristics. The filters with CPW pads have been fabricated in LTCC ($\varepsilon_r = 5.4$, $\tan \delta = 0.0015$) with a dielectric layer thickness of 100 μm and metal thickness of 9 μm. The overall size is 4.018 mm × 1.140 mm × 0.3 mm, including the CPW measurement pads. As shown in Fig. 4.6, the experimental and the simulated results agree very well. It can be easily observed that the insertion loss is <2.3 dB, the return loss >25.3 dB over the passband and the 3-dB bandwidth is about 1 GHz. The center frequency shift from 59.85 to 59.3 GHz can be attributed to the fabrication accuracy (vertical coupling overlap affected by the alignment between layers and layer thickness tolerance).

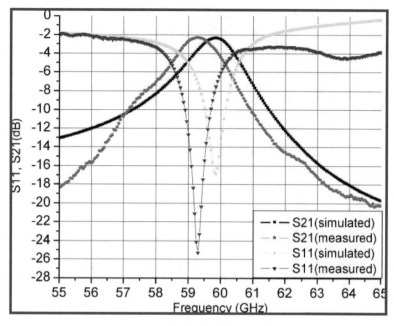

FIGURE 4.6: Measured and simulated S-parameters of 60 GHz slotted patch resonator.

4.1.2   Three and Five-Pole Resonator Filters

The next step for the easy and miniaturized realization of better rejection and selectivity would be the design of multistage filters. The presented example in this section deals with the design and fabrication of symmetrical three-pole and five-pole filters for intersatellite wideband applications that consist of, respectively, three and five capacitively gap-coupled single-mode resonators, as shown Fig. 4.7(a) and (b).

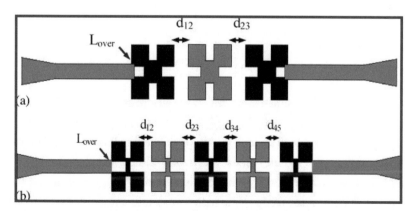

FIGURE 4.7: Top view of (a) three-pole slotted patch bandpass filter (b) five-pole slotted patch bandpass filter.

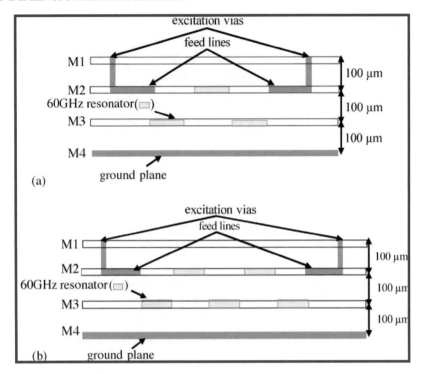

FIGURE 4.8: Side view of (a) three-pole slotted patch bandpass filter (b) five-pole slotted patch bandpass filter.

The first three-pole bandpass filter was developed for a center frequency of 59.6 GHz, 1dB insertion loss, 0.1 dB in band ripple, and 6.4% fractional bandwidth based on Chebyshev low-pass prototype filter. The design parameters, such as the external quality factors and the coupling coefficients, were

$$Q_{ext} = 15.4725$$

$$k_{12} = k_{23} = 0.06128.$$

To determine the physical dimensions, full-wave electromagnetic (EM) simulations (IE3D) were used to extract the coupling coefficients (k_{ii+1}, $i = 1$ or 2) and external quality factors (Q_{ext}) based on a simple graphical approach as described in [63]. Feeding lines and slotted patch resonators were alternately positioned on different metal layers (feeding lines, 2nd resonator: M2; 1st resonator, 3rd resonator: M3) as shown in Fig. 4.8(a),(b) to achieve strong k_{ii+1} between resonators as well as desired Q_{ext} between resonator and feeding line with a moderate sensitivity to the LTCC fabrication tolerances. The benefits of the multilayer filter topologies in terms of miniaturization can be easily observed in Fig. 4.8.

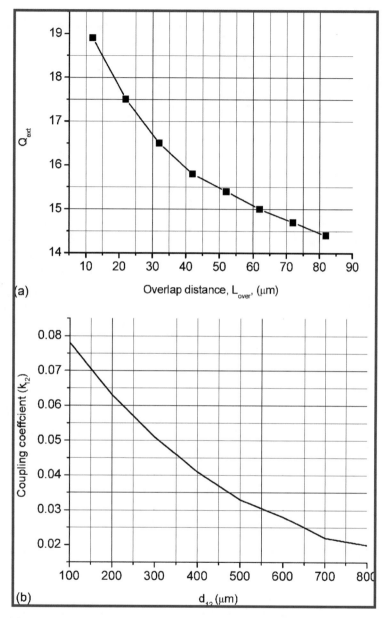

(a)

(b)

FIGURE 4.9: (a) External quality factor (Q_{ext}) evaluated as a function of overlap distance (L_{over}) (b) Coupling coefficient, k_{12}, as a function of coupling spacing (d_{12}) between 1st resonator and 2nd resonator.

Figure 4.9(a) shows the Q_{ext} evaluated as a function of overlap distance (L_{over}). A larger L_{over} results in a stronger input/output coupling and smaller Q_{ext}. Then, the required k_{ij} is obtained against the variation of distance [d_{ij} in Fig. 4.7(a)] for a fixed Q_{ext} at the input/output ports. Full-wave simulation was also employed to find two characteristic frequencies (f_{p1}, f_{p2}) that represent

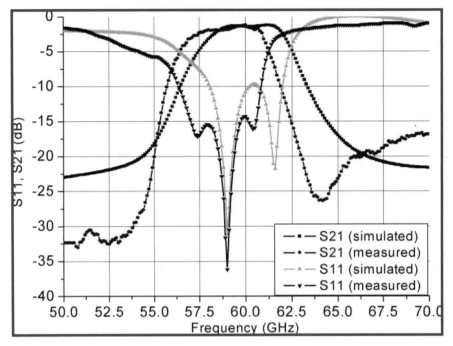

FIGURE 4.10: Measured and simulated S-parameters of three-pole slotted patch bandpass filter.

resonant frequencies of the coupled structure when an electrical wall or a magnetic wall, respectively, was inserted in the symmetrical plane of the coupled structure [63]. Characteristic frequencies were associated with the coupling between resonators as follows: $k = (f_{p2}^2 - f_{p1}^2)/(f_{p2}^2 + f_{p1}^2)$ [4]. The coupling spacing [d_{12} in Fig. 4.7(a)] between the first and second resonators for the required k_{12} was determined from Fig. 4.9(b). k_{23} and d_{23} are determined in the same way as k_{12} and d_{12} since the investigated filter is symmetrical around its center.

Figure 4.10 shows the comparison of the simulated and the measured S-parameters of the three-pole slotted patch filter. Good correlation is observed, and the filter exhibits an insertion loss <1.23 dB, the return loss >14.31 dB over passband, and the 3-dB bandwidth about 6.6% at center frequency 59.1 GHz. The selectivity on the high side of the passband is better than the EM simulation because an inherent attenuation pole occurs at the upper side. The latter is due to the fact that the space between fabricated nonadjacent resonators might be smaller than that in simulation so that stronger cross coupling might occur. In addition, the measured insertion loss is slightly higher than the theoretical result because of additional conductor loss and radiation loss from the feeding microstrip lines that cannot be de-embedded because of the nature of short, open, load, and thru (SOLT) calibration method. The dimensions of the fabricated filter are 5.855 mm × 1.140 mm × 0.3 mm with measurement pads.

A high-order filter design using five-slotted patches [Fig. 4.7(b)] and having very similar coupling scheme as the three-pole filter is also presented as an example for large (>3) number of filter stages. The Chebyshev prototype filter was designed for a center frequency of 61.5 GHz, 1.3 dB insertion loss, 0.1 dB band ripple, and 8.13% 3-dB bandwidth. The circuit parameters for this filter are:

$$Q_{ext} = 14.106$$

$$k_{12} = k_{45} = 0.0648$$

$$k_{23} = k_{34} = 0.0494$$

Figure 4.8(b) shows the side view of a five-pole slotted patch bandpass filter. The feeding lines and the open-circuit resonators have been inserted into the different metallization layers (feeding lines: 2nd and, 4th resonators M2; 1st, 3rd, and 5th resonator: M3) so that the spacing between adjacent resonators and the overlap between the feeding lines and the resonators work as the main parameters of the filter design to achieve the desired coupling coefficients and the external quality factor in a very miniaturized configuration. The filter layout parameters are $d_{12} = d_{45} \approx \lambda_{go}/16$, $d_{23} = d_{34} \approx \lambda_{go}/11$, $L_{over} \approx \lambda_{go}/26$ [Fig. 4.7(b)], where λ_{go} is the guided wavelength and the filter size is $7.925 \times 1.140 \times 0.3$ mm^3.

The measured insertion and reflection loss of the fabricated filter are compared with the simulated results in Fig. 4.11. The fabricated filter exhibits a center frequency of 59.15 GHz, an

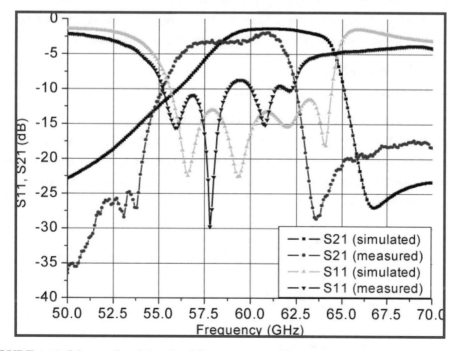

FIGURE 4.11: Measured and simulated S-parameters of five-pole slotted patch bandpass filter.

insertion loss of about 1.39 dB, and a 3-dB bandwidth of approximately 7.98%. These multipole filters can be used in the development of compact multi-pole duplexers. The difference between the measurement and simulation is attributed to the fabrication tolerances, as mentioned in the case of the three-pole bandpass filter.

4.2   QUASIELLIPTIC FILTER

Numerous researchers [63,64] have demonstrated narrow bandpass filters employing open-loop resonators for current mobile communication services at L- and S-bands. In this section, the design of a four-pole quasielliptic filter is presented as a filter solution for LTCC 60 GHz front-end module because it exhibits a superior skirt selectivity by providing one pair of transmission zeros at finite frequencies, enabling a performance between that of the Chebyshev and elliptical-function filters [63]. The very mature multilayer fabrication capabilities of LTCC ($\varepsilon_r = 7.1$, $\tan \delta = 0.0019$, metal layer thickness, 9 μm; number of layers, six; dielectric layer thickness, 53 μm; minimum metal line width and spacing, up to 75 μm) make it one of the leading competitive solutions to meet millimeter-wave design requirements in terms of physical dimensions [63] of the open-loop resonators ($\approx 0.2\lambda_g \times 0.2\lambda_g$), achieving a significant miniaturization because of relatively high ε_r, and spacing (≥ 80 μm) between adjacent resonators that determine the coupling coefficient of the filter function.

Figure 4.12(a) and (b) shows the top and cross-sectional views of a benchmarking microstrip quasielliptic bandpass filter, respectively. The filter was designed according to the filter synthesis proposed by Hong and Lancaster [63] to meet the following specifications:

1. Center frequency: 62 GHz;

2. Fractional bandwidth: 5.61% ($\sim$3.5 GHz);

3. Insertion loss: <3 dB (4) 35 dB rejection bandwidth: 7.4 GHz;

4. Its effective length [R_L in Fig. 4.12(a)] and width [R_W in Fig. 4.12(a)] has been optimized to be approximately $0.2\lambda_g$ using a full-wave simulator (IE3D) [63]. The design parameters, such as the coupling coefficients (C_{12}, C_{23}, C_{34}, C_{14}) and the Q_{ext} can be theoretically determined by the formulas [63].

$$Q_{ext} = \frac{g_1}{\text{FBW}}$$

$$C_{i,i+1} = C_{n-i,n-i+1} = \frac{\text{FBW}}{\sqrt{g_i g_{i+1}}} \qquad \text{for } i = 1 \text{ to } m - 1$$

$$C_{m,m+1} = \frac{\text{FBW} J_m}{g_m}$$

$$C_{m-1,m+2} = \frac{\text{FBW} J_{m-1}}{g_{m-1}}$$

(4.1)

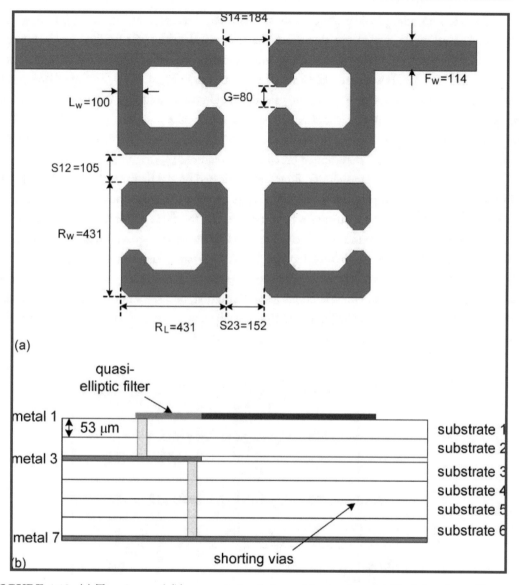

FIGURE 4.12: (a) Top view and (b) cross-sectional view of four-pole quasielliptic bandpass filter consisting of open-loop resonators fabricated on LTCC. All dimensions indicated in (a) are in μm.

where g_i is the element values of the low pass prototype, FBW is the fractional bandwidth, and J_i is the characteristic admittances of the filter. From (4.1) the design parameters of this bandpass filter are found:

$$C_{1,2} = C_{3,4} = 0.048, \quad C_{1,4} = 0.012, \quad C_{2,3} = 0.044, \quad Q_{ext} = 18$$

To determine the physical dimensions of the filter, numerous MoM-based full-wave EM simulations have to be carried out to extract the theoretical values of coupling coefficients and external quality factors [63]. The size of each square microstrip open-loop resonator is $431 \times 431\ \mu m^2$ [$R_W \times R_L$ in Fig. 4.12(a)] with the line width of $100\ \mu m$ [L_W in Fig.4. 12(a)] on the substrate. The coupling gaps [S23 and S14 in Fig. 4.12(a)] for the required $C_{2,3}$ and $C_{1,4}$ can be determined for the specific magnetic and electric coupling, respectively. The other coupling gaps [S12 and S34 in Fig. 4.12(a)] for $C_{1,2}$ and $C_{3,4}$ can be easily calculated for the mixed coupling. The tapered line position [T in Fig. 4.12(d)] is determined based on the required Q_{ext}.

One prototype of this quasielliptic filter was fabricated on the first metallization layer [metal 1 in Fig. 4.12(b)] that was placed two substrate layers ($\sim$106 μm) above the first ground plane on metal 3. That is the minimum substrate height to realize the 50 Ω microstrip feeding structure on LTCC substrate. This ground plane was connected to the second ground plane located on the backside of the substrate through shorting vias (pitch: 390 μm, diameter: 130 μm), as shown in Fig. 4.12(b) [65]. The four additional substrate layers [substrates 3–6 in Fig. 4.12(b)] were reserved for an integrated filter and antenna functions implementation, because antenna bandwidth requires higher substrate thickness than the filter, verifying the advantageous feature of the 3D modules that they can easily integrate additional or reconfigurable capabilities.

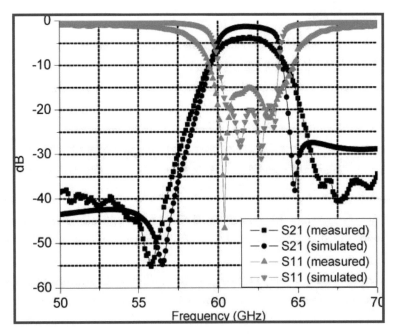

FIGURE 4.13: The comparison between measured and simulated S-parameters (S21 and S11) of the four-pole quasielliptic bandpass filter composed of open-loop resonators.

Figure 4.13 shows the comparison between the simulated and measured S-parameters of the bandpass filer. The filter exhibits an insertion loss <3.5 dB which is higher than the simulated values of <1.4 dB and a return loss >15 dB compared to a simulated value of <21.9 dB over the passpand. The loss discrepancy can be attributed to conductor loss caused by the strip edge profile and the quality of the edge definition of metal traces since the simulations assume a perfect definition of metal strips. Also, the metallization surface roughness may influence the ohmic loss because the skin depth in a metal conductor is very low at these high frequencies. The measurement shows a slightly decreased 3-dB fractional bandwidth of 5.46% (~3.4 GHz) at a center frequency of 62.3 GHz. The simulated results give a 3-dB bandwidth of 5.61% (~3.5 GHz) at a center frequency 62.35 GHz. The transmission zeros are observed within less than 5 GHz away from the cutoff frequency of the passband. The discrepancy of the zero positions between the measurement and the simulation can be attributed to the fabrication tolerance. However, the overall response of the measurement correlates very well with the simulation.

CHAPTER 5

Cavity-Type Integrated Passives

5.1 RECTANGULAR CAVITY RESONATOR

In numerous high-power microwave applications (e.g. remote sensing and radar), waveguide-based structures are commonly used due to their better power handling capability, although they are often bulky and heavy. In addition, this type of structures suffers from high metal loss due to the metallized walls, especially in the mm-wave frequency range, something that necessitates the modification of the conductor implementation for an easy 3D integration. The hereby presented cavity resonators are based on the theory of rectangular cavity resonators [62], built utilizing conducting planes as horizontal walls and via fences as sidewalls, as shown in Fig. 5.1. The size (d) and spacing (p) (see Fig. 5.1) of via posts are properly chosen to prevent electromagnetic field leakage and to achieve stop-band characteristic at the desired resonant frequency [27]. The resonant frequency of the TE_{mnl} mode is obtained by [62]

$$f_{res} = \frac{c}{2\pi\sqrt{\varepsilon_r}}\sqrt{\left(\frac{m\pi}{L}\right)^2 + \left(\frac{n\pi}{H}\right)^2 + \left(\frac{l\pi}{W}\right)^2} \tag{5.1}$$

where f_{res} is the resonant frequency, c the speed light in the free space, ε_r the dielectric constant, L the length of cavity, W the width of cavity, and H the height of the cavity. Using (5.1), the initial dimensions of the cavity with perfectly conducting walls are determined for a resonant frequency of 60 GHz for the TE_{101} dominant mode by simply indexing $m = 1$, $n = 0$, $l = 1$ and are optimized with a full-wave electromagnetic simulator ($L = 1.95$ mm, $W = 1.275$ mm, $H = 0.3$ mm). Then, the design parameters of the feeding structures are slightly modified to achieve the best performance in terms of low insertion loss and accurate resonant frequency.

To decrease the metal loss and enhance the quality factor, the vertical conducting walls are replaced by a lattice of via posts. In our case, we use Cassivi and Wu's expressions [66] to get the preliminary design values, and then the final dimensions of the cavity are fine tuned with the HFSS simulator. The spacing (p) between the via posts of the sidewalls is limited to less than half of the guided wavelength ($\lambda_g/2$) at the highest frequency of interest so that the radiation losses become negligible [27]. Also, it has been proven that smaller via sizes result in an overall size reduction of the cavity [27]. In our case, we used the minimum diameter of vias ($d = 130$ µm in Fig. 5.1) allowed by the LTCC design rules. Also, the spacing between the vias has been set to be the minimum via pitch (390 µm).

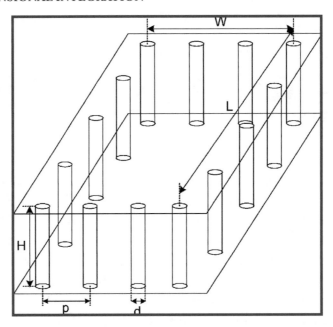

FIGURE 5.1: Cavity resonator utilizing via fences as sidewalls.

In the case of low external coupling, the unloaded unloaded Quality Factor, Q_u, is controlled by three loss mechanisms and defined by [61]

$$Q_u = \left(\frac{1}{Q_{\text{cond}}} + \frac{1}{Q_{\text{dielec}}} + \frac{1}{Q_{\text{rad}}} \right)^{-1} \qquad (5.2)$$

where Q_{cond}, Q_{dielec}, and Q_{rad} take into account the conductor loss from the horizontal plates (the metal loss of the horizontal plates dominates especially for thin dielectric thicknesses, H, such as 0.3 mm), the dielectric loss from the filling dielectrics, and the leakage loss through the via walls, respectively. Since the gap between the via posts is less than $\lambda_g/2$ at the highest frequency of interest, the leakage (radiation) loss can be negligible, as mentioned above, and the individual contribution of the two other quality factors can be obtained from [61]

$$Q_{\text{cond}} = \frac{(kWL)^3 H\eta}{2\pi^2 R_m (2W^3H + 2L^3H + W^3L + L^3W)} \qquad (5.3)$$

where k is the wave number in the resonator $((2\pi f_{\text{res}}(\varepsilon_r)^{1/2})/c)$, R_m is the surface resistance of the cavity ground planes $((\pi f_{\text{res}}\mu/\sigma)^{1/2})$, η is the wave impedance of the LTCC resonator filling, L, W, and H are the length, width, and height of the cavity resonator, respectively and

$$Q_{\text{dielec}} = \frac{1}{\tan \delta} \qquad (5.4)$$

where $\tan \delta$ is the loss tangent ($=0.0015$) of the LTCC substrate. The quality factor [Eqs. (5.2)–(5.4)] of a rectangular cavity can be used effectively in the cavity using via-array sidewalls, which almost match the performance of the PECs [26,29].

The loaded quality factor (Q_l) can be obtained by adding the losses (Q_{ext}) of the external excitation circuit to the Q_u as expressed in [61]

$$Q_l = \left(\frac{1}{Q_u} + \frac{1}{Q_{ext}} \right)^{-1} \tag{5.5}$$

The theoretical values of Q can be extracted from the simulated performances of a weakly coupled cavity resonator using the following equations [61]:

$$Q_l = \frac{f_{res}}{\Delta f} \tag{5.6}$$

$$S21(dB) = 20 \log_{10} \left(\frac{Q_l}{Q_{ext}} \right) \tag{5.7}$$

$$Q_u = \left(\frac{1}{Q_l} - \frac{1}{Q_{ext}} \right)^{-1} \tag{5.8}$$

where Δf is the 3-dB bandwidth. The weak external coupling allows for the verification of Q_u of the cavity resonator as Q_u approaches Q_l with the weak external coupling as described in (5.8). Also the weak coupling abates the sensitivity of the measurement on the amplitude of S21. Using the above definitions, a weakly coupled cavity resonator (S21~20 dB) has been separately investigated in HFSS and exhibits a Q_u of 367 at 59.8 GHz compared to the theoretical Q_u of 372 at 60 GHz from (6)–(8). All fabricated resonators were measured using the Agilent 8510C Network Analyzer and Cascade Microtech probe station with 250 μm pitch air coplanar probes. A standard short-open-load-through (SOLT) method was employed for calibration.

5.2   THREE-POLE CAVITY FILTERS

The next topology covered in this chapter has to do with three-pole filters using via walls for 60 GHz WLAN narrowband (~1 GHz) applications that consist of three coupled cavity resonators [cavity 1, cavity 2, cavity 3 in Fig. 5.2(b)]. The three-dimensional (3D) overview and side view are illustrated in Fig. 5.2(a) and (b), respectively. The three-pole bandpass filter based on a Chebyshev lowpass prototype filter is developed for a center frequency of 60 GHz, <3 dB insertion loss, 0.1 dB in band ripple and 1.67% fractional bandwidth.

To meet design specifications, the cavity height [H in Fig. 5.2(a)] was set to 0.5 mm (five substrate layers) to achieve a higher Q_u and consequently to obtain narrower bandwidth. The cavity resonator with 0.5 mm height has been fabricated in LTCC and measured. The comparison between the simulation and the measurement is shown in Fig. 5.3. An insertion loss of 1.24 dB at the center

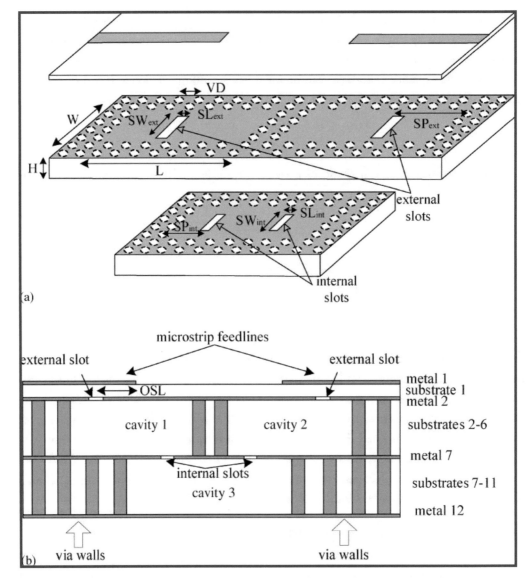

FIGURE 5.2: LTCC three-pole cavity bandpass filter employing slot excitation with an open stub: (a) 3D overview and (b) side view of the proposed filter.

frequency of 59.2 GHz and a narrow bandwidth of 1.35% (~0.8 GHz) has been measured. The theoretical Q_u yields 426, and it is very close to the simulated Q_u of 424 from a weakly coupled cavity in HFSS.

After verifying the experimental performance of a single cavity resonator, the external coupling and the interresonator coupling are considered for the three-pole filter design. These factors are very important in the design of multi-cavity (multi-pole) filters.

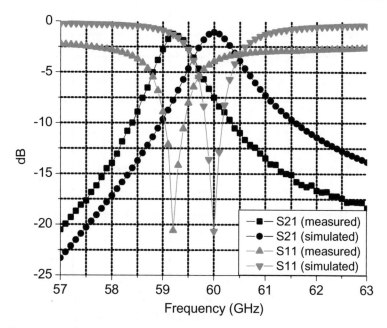

FIGURE 5.3: Comparison between measured and simulated S-parameters (S11 and S21) of 0.5-mm height cavity resonator using slot excitation with an open stub.

Firstly, Q_{ext} can be defined from the specifications as follows [67]:

$$Q_{ext} = \frac{g_i g_{i+1} f_{res}}{FBW} \qquad (5.9)$$

where g_i are the element values of the low pass prototype, f_{res} is the resonant frequency, and FBW is the fractional bandwidth of the filter. The input and output Q_{ext} were calculated to be 61.89. The position and size of the external slots [Fig. 5.2(a)] are the main parameters to achieve the desired Q_{ext}. The slots have been positioned at a quarter of the cavity length ($L/4$) and their length has been fixed to $\lambda_g/4$ [$SL_{ext} \approx \lambda_g/4$ in Fig. 5.2(a)]. Then, Q_{ext} [shown in Fig. 5.4(a)] has been (using full-wave simulations) evaluated as a function of the external slot width [SW_{ext} in Fig. 5.2(a)] based on the following relationship [26]

$$Q_{ext} = \frac{f_{res}}{\Delta f_{\pm 90°}} \qquad (5.10)$$

where $\Delta f_{\pm 90°}$ is the frequency difference between $\pm 90°$ phase response of S11.

Secondly, the interresonator coupling coefficients (k_{jj+1}) between the vertically adjacent resonators is determined by [67]

$$k_{j,j+1} = \frac{BW}{f_{res}} \sqrt{\frac{1}{g_j g_{j+1}}} \qquad (5.11)$$

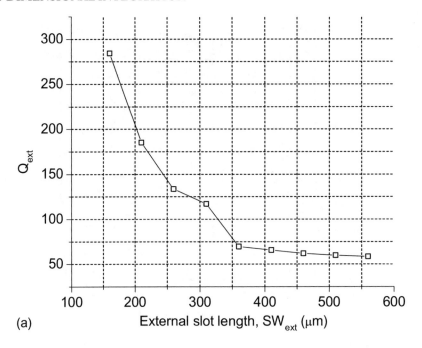

(a)

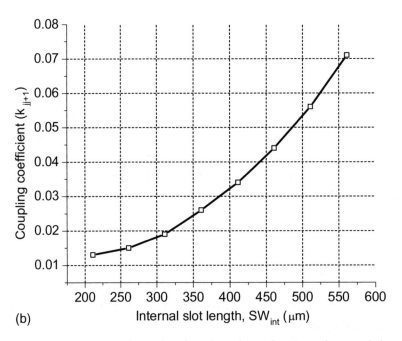

(b)

FIGURE 5.4: (a) External quality factor (Q_{ext}) evaluated as a function of external slot width (SW_{ext}). (b) Interresonator coupling coefficient (k_{jj+1}) as a function of internal slot width (SW_{int}).

where $j = 1$ or 2 because of the symmetrical nature of the filter. k_{jj+1} was calculated to be 0.0153. To extract the desired k_{jj+1}, the size of internal slots [Fig. 5.2(a)] is optimized using full-wave simulations to find the two characteristic frequencies (f_{p1}, f_{p2}) that are the frequencies of the peaks in the transmission response of the coupled structure when an electric wall or magnetic wall, respectively, is inserted in the symmetrical plane [67]. Then, k_{jj+1} can be determined by measuring the amount that the two characteristic frequencies deviate from the resonant frequency. The relationship between k_{jj+1} and the characteristic frequencies (f_{p1}, f_{p2}) is defined as follows: [67].(15)

$$k_{jj+1} = \frac{f_{p2}^2 - f_{p1}^2}{f_{p2}^2 + f_{p1}^2} \qquad (5.12)$$

Based on the above theory, the physical dimensions of internal slots can be determined by using a simple graphical approach displaying two distinct peaks of character frequencies for a fixed Q_{ext}. Figure 5.4(b) shows the graphical relationship between k_{jj+1} and internal slot width [SW$_{int}$ in Fig. 5.2(a)] variation with the fixed slot length [SL$_{int} \approx \lambda_g/4$ in Fig. 5.2(a)]. SW$_{int}$ was determined to be 0.261 mm corresponding to the required $k_{jj+1}(\approx 0.0153)$ from Fig. 5.4(b). After determining the initial dimensions of the external/internal slots, the other design parameters such the cavity length and width [L and W in Fig. 5.2(a)] using via walls are determined under the design guidelines described in Section 5.1.

The initial dimensions of the external/internal slot widths are set up as optimal variables and fine-tuned to achieve the desired frequency response using HFSS simulators. The summary of all design parameters for the three-pole filter is given in Table 5.1. Figure 5.5(a) and (b) shows the comparison between the simulated and the measured S-parameters of the

TABLE 5.1: Design parameters of three-pole cavity filter using an open stub.

DESIGN PARAMETERS	DIMENSIONS (MM)
Effective cavity resonator (L × W × H)	1.95 × 1.32 × 0.5
External slot position (SP$_{ext}$)	0.4125
External slot (SL$_{EXT}$ × SW$_{ext}$)	0.46 × 0.538
Internal slot position (SP$_{int}$)	0.3915
Internal slot (SL$_{INT}$ × SW$_{int}$)	0.261 × 0.4
Open stub length (OSL)	0.538
Via spacing	0.39
Via diameter	0.13
Via rows	3

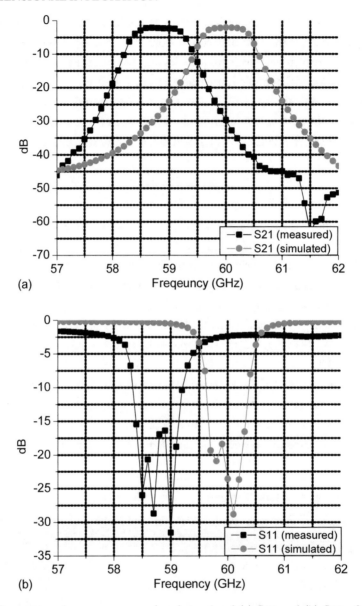

FIGURE 5.5: Comparison between measured and simulated (a) S21 and (b) S11 of three-pole cavity bandpass filter using slot excitation with an open stub.

bandpass filter. In the measurements, the parasitic effects from the I/O open pads were de-embedded with the aid of WinCal 3.0 software. The filter exhibits an insertion loss <2.14 dB which is slightly higher than the simulated value of <2.08 dB, and a return loss >16.39 dB compared to a simulated value >18.37 dB over the pass band, as shown in Fig. 5.5(a) and (b), respectively. In Fig. 5.5(a), the measurement shows a slightly increased 3 dB fractional bandwidth of about 1.53%

($\approx$0.9 GHz) at a center frequency 58.7 GHz. The simulated results give a 3-dB bandwidth of 1.47% ($\approx$0.88 GHz) at a center frequency 60 GHz. The center frequency downshift can be attributed to the fabrication accuracy issues, such as slot positioning affected by the alignment between layers, layer thickness tolerance, and higher dielectric constant at this high frequency range (55–65 GHz) than 5.4 that is the relative permittivity at 35 GHz. The overall response of the measurement is in excellent agreement with the simulation except a frequency shift of 1.3 GHz ($\sim$2%).

5.3   VERTICALLY STACKED CAVITY FILTERS AND DUPLEXERS

An even more compact filter configuration, that takes better advantage of the third (vertical) dimension and further reduces its horizontal area is realized by stacking vertically numerous cavities. An example of this approach is the vertically stacked cavity bandpass filter that is presented in this section. This topology is designed in a way that allows for its easy integration with a V-band multilayer module due to its compactness and its 3D interconnect feature allowing for its use as a duplexer between the active devices on the top of the LTCC board and the antenna integrated on the backside. High level of compactness can be achieved by vertically stacking three identical cavity resonators with the microstrip feedlines vertically coupled through rectangular slots etched on the input and output resonators. The presented benchmarking topologies were fabricated in an LTCC. The relative permittivity (ε_r) of the substrate is 5.4 and its loss tangent (tan δ) is 0.0015. The dielectric thickness per layer is 100 μm, and the metal thickness is 9 μm. The resistivity of metal (silver trace) is determined to be 2.7×10^{-8} Ω m.

5.3.1   Design of Cavity Resonator

The cavity resonator that is the most fundamental component of the cavity filter is built based on the conventional rectangular cavity resonator approach investigated in Section 5.1. The cavity resonator shown in Fig. 5.6 consists of one LTCC cavity, two microstrip lines for input and output, and two vertically coupling slots etched on the ground planes of the cavity. The resonant frequency of the fundamental TE_{101} mode can be determined by (5.1) and its value around 60.25 GHz establishes the initial dimensions of the cavity resonator enclosed by perfectly conducting walls. For the purpose of compactness, the height (H) is determined to be 0.1 mm (one substrate layer). Then, the vertical conducting walls are replaced by double rows of via posts that are sufficient to suppress the field leakage and to enhance the Q. In addition, the size and spacing of via posts are properly chosen to prevent electromagnetic field leakage and to achieve stop-band characteristic at the desired resonant frequency according to the guidelines specified in Section 5.1. In the presented example, the minimum value of center-to-center vias spacing ($p = 390$ μm) and the minimum value of via diameter of the LTCC design rules ($d = 130$ μm) are used (see Fig. 5.6). The final dimensions of the via-based cavity are determined by using a tuning analysis of HFSS full-wave simulator ($L = 1.95$ mm, $W = 1.275$ mm, $H = 0.1$ mm).

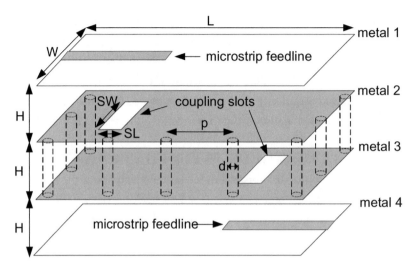

FIGURE 5.6: 3D overview of LTCC cavity resonator employing slot excitation with microstrip feedlines on the different metal layers (metals 1 and 4).

With the cavity size determined, microstrip lines are utilized as the feeding structure to excite the cavity via coupling slots that couple energy magnetically from the microstrip lines into the cavity. For a preliminary testing of the vertical intercoupling of three-pole cavity bandpass filter, the input and output feedlines are placed on metals 1 and 4, respectively; the coupling coefficient can be controlled by the location and size of the coupling slots etched on metals 2 and 3 (see Fig. 5.6).

To accurately estimate the Q_u, the weakly coupled cavity resonator [68] with a relatively small value of the slot length is implemented in HFSS simulator (SL in Fig. 5.6). Q_u can be extracted from the Q_{ext} and the Q_l using (5.5)–(5.8). The simulated value of Q_u was calculated to be 623 at 60.25 GHz.

5.3.2   Design of Three-Pole Cavity Bandpass Filter

As a demonstration of the above design approach, a vertically stacked LTCC three-pole cavity bandpass filter is developed for 3D integrated 59–64 GHz industrial, scientific, and medical (ISM) band transceiver front-end modules. The center frequencies of 60.25 GHz and 62.75 GHz in the band are selected for the Rx channel and the Tx channel, respectively.

First, the cavity bandpass filter for the Rx channel selection is designed with a 60.25 GHz center frequency, a <3 dB insertion loss, a 0.1 dB ripple, and a 4.15% (≈2.5 GHz) fractional bandwidth based on a Chebyshev lowpass prototype. The filter schematic is implemented with 10 substrate layers of LTCC tape. Its 3D overview, side view, top view of the feeding structure, and interresonator coupling structure are illustrated in Fig. 5.7(a)–(d), respectively. The top five substrate layers [substrates 1–5 in Fig. 5.7(b)] are occupied by the Rx filters, and the remaining layers are reserved

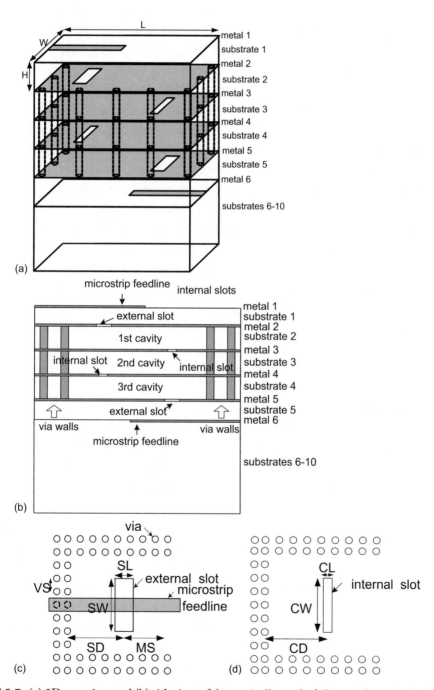

FIGURE 5.7: (a) 3D overview and (b) sideview of the vertically stacked three-pole cavity bandpass filter. Top view of the (c) feeding structure and (d) interresonator coupling structure.

for the antenna and the RF active devices, that could be integrated into front-end modules. The microstrip lines on metals 1 and 6 are utilized as the feeding structure to excite the 1st and 3rd cavities, respectively. Three identical cavity resonators [1st cavity, 2nd cavity, 3rd cavity in Fig. 5.7(b)], designed in Section 5.3.1, are vertically stacked and coupled through slots to achieve the desired frequency response with high level of compactness. This filter is also an effective solution to connect the active devices on the top of the LTCC board and the antenna integrated on the backside.

Two external slots [Fig. 5.7(a)] on metal layers 2 and 5 are dedicated to magnetically couple the energy from the I/O microstrip lines into the 1st and 3rd cavity resonators, respectively.

To maximize magnetic coupling by maximizing the current, the microstrip feedlines are terminated with a $\lambda_g/4$ open stub beyond the center of each external slot. The fringing field generated by an open-end discontinuity can be modeled by an equivalent length of transmission line determined to be about $\lambda_g/20$. Therefore, the optimum length of the stub is approximately $\lambda_g/5$ [MS in Fig. 5.7(c)]. The position and size of the external slots are the main design parameters to provide the necessary Q_{ext}. The external quality factor (Q_{ext}) that controls the insertion loss and ripple over the passband can be defined by (5.9).

The calculated Q_{ext} is 24.86. The external slot is initially positioned at $L/4$ from the edge of the cavity, and the width [SW in Fig. 5.7(c)] of the slot is fixed to $\lambda_g/4$. Then, the length [SL in Fig. 5.7(c)] of the slot is tuned until the simulated Q_{ext} converges to the prototype requirement. Figure 5.8 shows the relationship between the length variation of the external slots and the Q_{ext} extracted from the simulation using (5.10).

The latter internal slots on metals 3 and 4 [Fig. 5.7(b)] are employed to couple energy from the 1st and 3rd cavity resonators into the 2nd resonator, and their design procedure is similar to that of the external slots. The internal slots are located at a quarter of the cavity length from the sides. The desired interresonator coupling coefficients ($k_{12} = k_{23} = 0.0381$) are obtained by (5.11). This desired prototype $k_{j,j+1}$ can be physically realized by varying the slot length [CL in Fig. 5.7(d)] with a fixed slot width [CW $\approx \lambda_g/4$ in Fig. 5.7(d)]. Full-wave simulations are employed to find the two characteristic frequencies (f_{p1}, f_{p2}) that are the resonant frequencies in the transmission response of the coupled structure [67]; its plot versus frequency is shown in Fig. 5.9(a). These characteristic frequencies are associated with the interresonator coupling between the cavity resonators according to (5.12).

Figure 5.9(b) shows the internal coupling as a function of the internal slot length [CL in Fig. 5.7(d)]. By adjusting the slot length, the optimal size of an internal slot can be determined for a given prototype value. Using the initial dimensions of the external (SW, SL) and internal slot (CW, CL) size as the design variables, we optimized the design variables to realize the desired frequency response. The design can be fine-tuned after considering the minimum and maximum of the fabrication tolerances. Then, the final values that match the desired frequency response can be determined.

To allow the wafer characterization using coplanar probes, the input and output probe pads have to be on the same layer, which requires an embedded microstrip line to CPW vertical transition

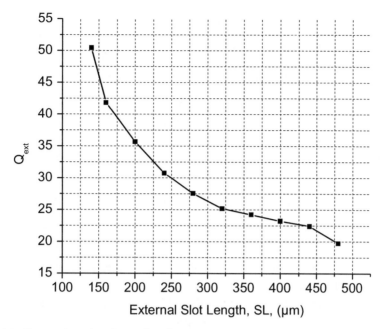

FIGURE 5.8: External quality factor (Q_{ext}) evaluated as a function of external slot length (SL).

at port 2. The vertical transition consists of five stacked signal vias penetrating through circular apertures [see Fig. 5.10(a)] on the ground planes (metals 2, 3, 4, and 5) and connecting an embedded microstrip line on metal 6 to a CPW measurement pads on metal 1. In order to match to the 50 Ω feedlines, the diameter of the circular apertures is optimized to be 0.57 mm for a signal via diameter of 130 μm, while the width of the microstrip line tapers out as it approaches the overlying CPW. Also, eight shielding vias (two connecting, metals 1 (CPW ground planes) to 5, six connecting, metals 2 to 5) are positioned around the apertures to achieve an optimum coaxial effect [69]. The number of shielding vias is determined with regard to the LTCC design rules.

The filters including CPW pads and a vertical transition were fabricated in LTCC. And measured on a HP8510C Vector Network Analyzer using SOLT calibration. Figure 5.10(a) depicts the 3D overview of the complete structure that was simulated. The "Wincal" software gives us the ability to de-embed capacitance effects of CPW open pads and inductive effects of short pads from the measured S-parameters so that the loading shift effect could be negligible. Figure 5.10(b) shows the photograph of the fabricated filter with CPW pads and a transition whose size is $5.60 \times 3.17 \times 1$ mm^3. The cavity size is determined to be $1.95 \times 1.284 \times 0.1$ mm^3 [$L \times W \times H$ in Fig. 5.6].

Figure 5.11(a) shows a comparison between the simulated and the measured S-parameters of the three-pole vertically stacked bandpass filter. The filter exhibits an insertion loss <2.37 dB, which is higher than the simulated value of <1.87 dB. The main source of this discrepancy might be caused by the radiation loss from the "thru" line that could not be de-embedded because of the

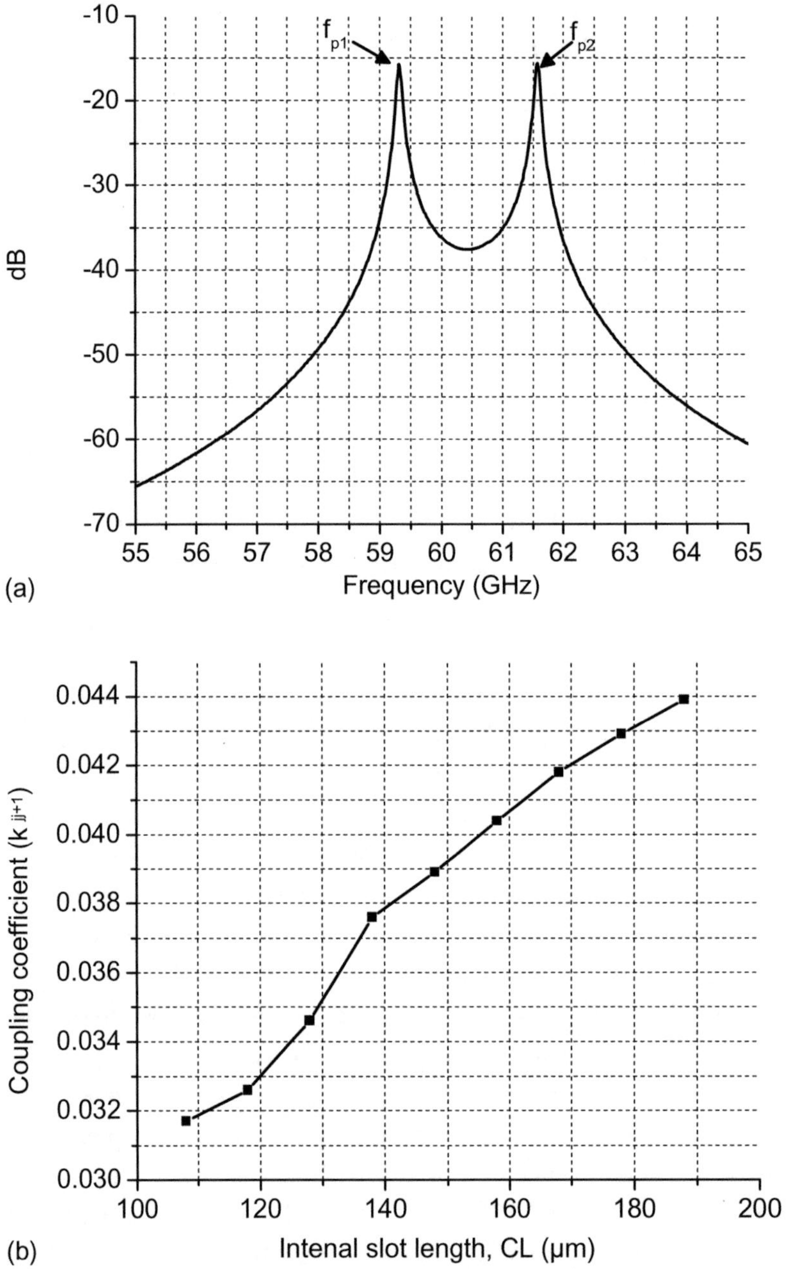

FIGURE 5.9: (a) Two characteristic frequencies (f_{p1}, f_{p2}) of the coupled cavities to calculate the internal coupling coefficients (k_{jj+1}). (b) Interresonator coupling coefficient (k_{jj+1}) as a function of internal slot length (CL).

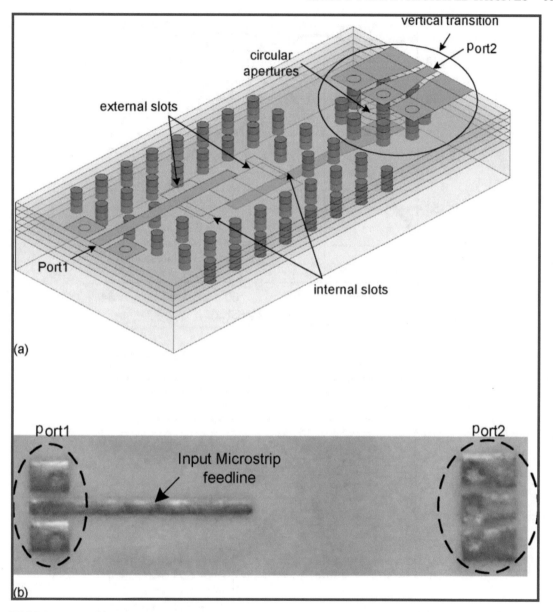

FIGURE 5.10: (a) 3D overview of vertically stacked three-pole cavity bandpass filter with CPW pads and vertical transitions. (b) Photograph of the cavity bandpass filter fabricated on LTCC.

nature of SOLT calibration. The filter exhibits a 3-dB bandwidth about 3.5% ($\approx$2 GHz) comparable to the simulated 3.82% ($\approx$2.3 GHz). The narrower bandwidth in measurements might be due to the fabrication accuracy of the slot design that has been optimized for the original resonant frequencies and not for the shifted frequencies.

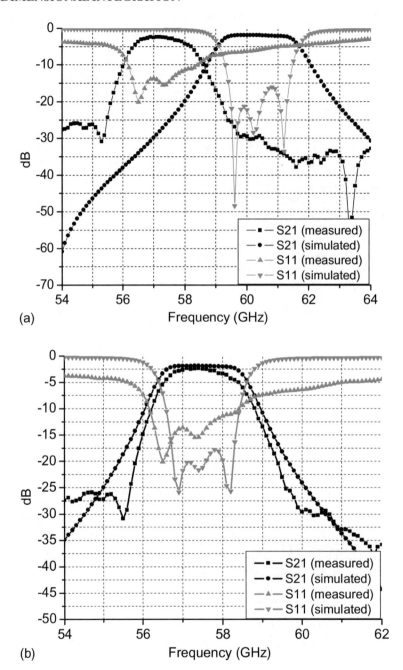

FIGURE 5.11: Comparison between measured and simulated S-parameters (S11 & S21) of Rx three-pole cavity band filter. (a) Measurement versus simulation with $\varepsilon_r = 5.4$ and originally designed cavity size ($1.95 \times 1.284 \times 0.1 \text{ mm}^3$). (b) Measurement versus simulation with $\varepsilon_r = 5.5$ and modified cavity size ($2.048 \times 1.348 \times 0.1 \text{ mm}^3$).

The center frequency shift from 60.2 GHz to 57.5 GHz might be attributed to the dielectric constant variation at these high frequencies and the fabrication accuracy of vias positioning caused by XY shrinkage. The HFSS simulation is re-performed in terms of two aspects. (1) The dielectric constant of 5.4 was extracted using cavity resonator characterization techniques [13] at 35 GHz. The dielectric constant is expected to increase to 5.5 across 55–65 GHz [21]. (2) The tolerance of XY shrinkage is expected to be ±15%. XY shrinkage specification was released after design tape out; thus, we could not have accounted for it at the design stage, and it can significantly affect the via positioning that is the major factor to determine the resonant frequency of a cavity filter. From our investigation, the averaged relative permittivity was evaluated to be 5.5 across 55–65 GHz [13], and the cavity size was modified to 2.048 × 1.348 × 0.1 mm^3 with 5% of XY shrinkage effect. The exact coincidence between the measured center frequency (57.5 GHz) and the simulated (57.5 GHz) is observed in Fig. 5.11(b). All design parameters for the modified Rx filter are summarized in Table 5.2.

The same techniques were applied to the design of the cavity bandpass filter for the Tx channel (61.5–64 GHz). The Chebyshev prototype filter was designed for a center frequency of 62.75 GHz, a <3 dB insertion loss, a 0.1 dB band ripple and a 3.98% 3-dB bandwidth. To meet the specified center frequency specifications, the cavity width (W) was decreased. Then the cavity size was determined to be 1.95 × 1.206 × 0.1 [$L \times W \times H$ in Fig. 5.7(a)] mm^3. The external and internal coupling slot sizes are used as the main design parameters to obtain the desired external quality factors and coupling coefficients, respectively.

The measured results of the Tx filter exhibit an insertion loss of 2.39 dB with a 3-dB bandwidth of 3.33% (∼2 GHz) at the center frequency of 59.9 GHz. The center frequency is downshifted

TABLE 5.2: Design parameters of cavity resonators.

DESIGN PARAMETERS	1ST CHANNEL (R_X)	2ND CHANNEL (T_X)
Cavity length (L)	2.048	2.048
Cavity width (W)	1.348	1.266
Cavity height (H)	0.100	0.100
External slot width (SW)	0.628	0.621
External slot length (SL)	0.460	0.460
External slot position (SD)	0.417	0.417
Internal slot width (CW)	0.558	0.551
Internal slot length (CL)	0.138	0.138
Internal slot position (CD)	0.417	0.417
Open stub length (MS)	0.571	0.571

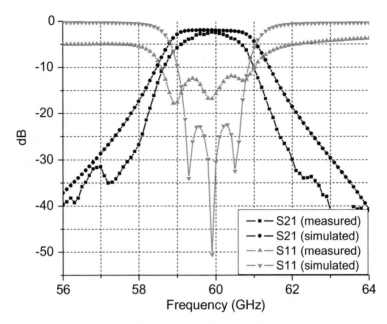

FIGURE 5.12: Comparison between measured and simulated S-parameters (S11 & S21) of Tx three-pole cavity band filter (simulation with $\varepsilon_r = 5.5$ and modified cavity size ($2.048 \times 1.266 \times 0.1$ mm^3) versus measurement).

approximately 2.72 GHz similarly to the Rx filter. A new theoretical simulation was performed with $\varepsilon_r = 5.5$ and the 5% increase in the volume of cavity ($2.048 \times 1.266 \times 0.1$ mm^3), and the measured and simulated results are presented in Fig. 5.12. The simulation showed a minimum insertion loss of 1.97 dB with a slightly increased 3-dB bandwidth of 4% ($\sim$2.4 GHz). The center frequency of the simulated filter was 59.9 GHz. The center frequency shift is consistent for both Tx and Rx devices due to their fabrication utilizing the same LTCC process. It has to be noted that the above two factors (dielectric constant frequency variation and dimension modification due to the co-firing) are the major issues that have to be considered in practical 3D cavity topologies in LTCC, especially in the mm-wave frequency range. All design parameters for the modified Tx filter are summarized in Table 5.2.

5.4   CAVITY-BASED DUAL-MODE FILTERS (HIGH-FREQUENCY SELECTIVITY)

In the previous sections, we developed single-mode cavity resonators and three-pole bandpass filters by adopting the vertical deployment of three single-mode cavity resonators. However, these single-mode devices could not satisfy optimum frequency selectivity. To achieve this selectivity with a compact size and reduced weight, dual-mode dielectric rectangular [70–77] and circular waveguide filters [78–81] have been proposed. The developed waveguide dual-mode filters make use of the coupling of two orthogonal modes generated from tuning screws [70–73,78,80,81], rectangular ridges [76,77], or the offsets of the feeding structure [74,75,79]. Multipole, dual-mode cavity filters have been realized for higher frequency selectivity through the coupling between modes in adjacent dual-mode, single waveguide resonators using a cross slot [72,73,78,79,81] or rectangular irises [76–80] or rectangular waveguides [75], [617]. However, these techniques not only impose a very heavy numerical burden to the modal characterization of waveguides because of the large number of evanescent modes, but also are not applicable to LTCC multilayer processes because of the fabrication limitations against a solid metal wall.

In this section, we expand previous work to a new class of 3D V-band dual-mode cavity filters and vertically stacked multipole filters using LTCC technologies, which enable a variety of quasielliptic responses by controlling the locations of transmission zeros. In Section 5.4.1, a dual-mode single cavity filter is developed for Rx and Tx channels as a complete filter solution in the design of V-band transceiver front-end modules. The appearance and elimination of transmission zeros have been analyzed through multipath coupling diagrams and lumped element models consisting of an intercoupling through the offset of feeding structures and a cross coupling by source-to-load spacings. To extend this approach to the design of multipole cavity filters, the vertically stacked arrangement of two dual-mode cavities is reported for the first time ever in Section 5.4.1. The presynthesized dual-mode single cavity filters are stacked with two different coupling slots (rectangular and cross) between the two cavities. The feasibility of realizing a multipole filter has been validated with the experimental data.

5.4.1   Dual-Mode Cavity Filters

5.4.1.1 Single Dual-Mode Cavity Resonator. The square-shaped cavity resonator is first designed at a center frequency of 63 GHz to exhibit a degenerate resonance of two orthogonal modes (TE_{102} and TE_{201}), characteristic of the dual-mode operation. LTCC multilayer substrates have been used for the fabrication, and their properties are as follows: The ε_r is 7.1, tan δ is 0.0017, the dielectric layer thickness is 53 μm per layer for a total of 5 layers, the metal thickness is 9 μm, and the resistivity of the metal (silver trace) is 2.7×10^{-8} Ωm. Figure 5.13(a) and (b) shows the 3D overview and the top view of the proposed structure, respectively. The dual-mode cavity resonator consists of one cavity occupying two substrate layers S2 and S3, the I/O microstrip feedlines on M1 and the two coupling slots etched on the top ground plane, M2 of the cavity. The microstrip lines are terminated with a physical short circuit realized by a metallic via (throughout S1) to maximize the magnetic coupling through the slots. In order to determine the effective length, L, and width, W, in Fig. 5.13(b) of the cavity resonator providing two orthogonal modes of TE_{mnl} and TE_{pqr}, both modes are designated to resonate at the same frequency using the conventional resonant frequency equation of the rectangular waveguide cavity.

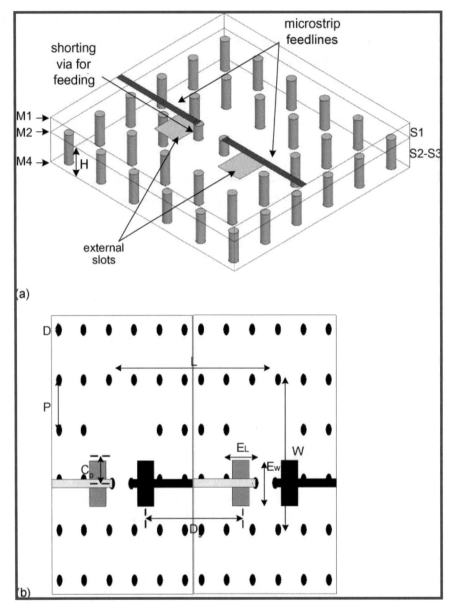

FIGURE 5.13: (a) 3D overview and (b) top view of a quasielliptic dual-mode single cavity filter.

The final dimensions of the cavity resonator using via fences as vertical walls are determined to be $2.06 \times 2.06 \times 0.106$ mm^3 in order to resonate at 63 GHz. The size and spacing of the via posts are properly chosen according to the LTCC design rules, such as the minimum value of center-to-center vias spacing p in Fig. 5.13(b) of 390 μm and the minimum value of via diameter d in Fig. 5.13(b) of 145 μm.

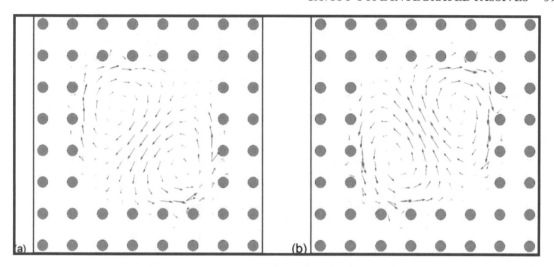

FIGURE 5.14: Magnetic vector of the (a) odd mode and (b) even mode.

5.4.1.2 Internal Coupling. The centerline offset, C_o, in Fig. 5.13(b) between the feeding structure and cavity position is one of major factors in realization of the dual-mode operation and controlling the mutual internal coupling of the modes, hence providing transmission zeros at the desired positions for a high selectivity. When the I/O slots are centered at the cavity interface ($C_o = 0$ mm), only the TE_{102} mode is excited so that the transmission zeros do not exist. However, when a transverse offset, C_o, is applied to the position of the I/O feeding structure, the additional mode, TE_{201}, mode is excited. This mode degeneration can be used to realize dual-mode filters.

The basis modes are defined as even and odd mode, respectively [28], (by vectorial addition and subtraction of TE_{102} and TE_{201} modes) and the magnetic vectors of these modes calculated using HFSS simulation software are displayed in Fig. 5.14. The resonant frequencies (f_e: even mode, f_o: odd mode) are associated with the intercoupling coefficient according to the definition of the ratio of the coupled energy to the stored energy of an uncoupled single resonator [82]. The value of f_e (f_o) can be derived from a symmetric structure by placing a prefect electric conductor (a perfect magnetic conductor) on the plane of the symmetry. Figure 5.15 displays the internal coupling coefficient as a function of the variation of the centerline offset C_o.

5.4.1.3 External Coupling. The I/O external slots on the top ground plane of the cavity are designed in a way that optimizes the magnetic excitation of the cavity from the 50 Ω microstrip lines. The accurate design of the external coupling slots that is directly related to the external quality factor, Q_{ext}, is a key issue to achieve a high-Q cavity resonator. The Q_{ext} corresponds to the resistance and the reactance and can be controlled by the position and size of the coupling slots. In order to investigate

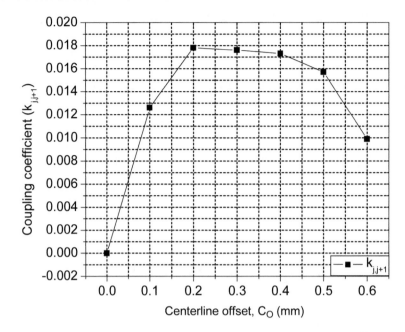

FIGURE 5.15: Internal coupling coefficient k_{jj+1} as a function of the centerline offset C_o of the feeding structures.

how the slot size affects the Q_{ext}, the external slots are initially placed at a quarter of the cavity length from the (front and back) edge of the cavity, and the slot length is varied with respect to the fixed slot width ($\sim\lambda_g/4$). The issues related to the distance between external slots (D_s in Fig. 5.13) will be discussed in detail in Section 5.4.1.4. Both single-mode case ($C_o = 0$ mm) and dual-mode case ($C_o = 0.6$ mm) were tested. In the single mode case, the Q_{ext} can be determined by the relation [67] between the resonant frequency and the frequencies where a $\pm 90°$ phase response in S11 parameter is exhibited. However, in the dual-mode case, the external coupling factor is directly related to the internal coupling coefficient according to the analytical equation [83]

$$Q_{ext} = \frac{1}{\sqrt{k_{1,2}^2 - k_{(1,2)wo}^2}} \tag{5.13}$$

where $k_{1,2}$ is the coupling coefficient of the dual-mode resonator with an external circuit and $k_{(1,2)wo}$ is the coupling coefficient of the dual-mode resonator without an external circuit.

Figure 5.16 shows the relationship between the length variation of the external slots E_L and the Q_{ext} from the simulation when the feeding structure is placed at 0.6 mm away from the center of the cavity ($C_o = 0.6$ mm). A larger E_L results in smaller Q_{ext} that is interpreted as a stronger external coupling.

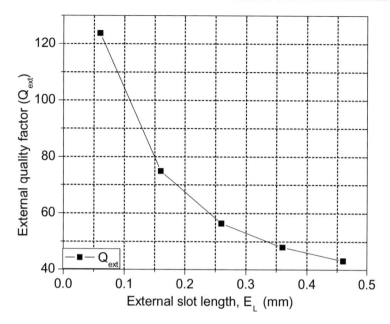

FIGURE 5.16: External quality factor Q_{ext} evaluated as a function of external slot length E_L.

5.4.1.4 Transmission Zero. In this section, the dual-mode filter realization with transmission zeros for high selectivity will be discussed. The equivalent circuit model of the proposed dual-mode cavity filter is shown in Fig. 5.17(a).

The filter consists of major four sections: (1) a pair of LC resonators that represent each of the degenerate dual modes in the cavity resonator, (2) mutual internal electric coupling, M, between a pair of parallel LC resonators, (3) external magnetic coupling, L_{ext} from each of the I/O external slots, and (4) magnetic cross coupling, L_c, representing the parasitic source to load coupling associated with the perturbed electric fields in the cavity [84]. A pair of transmission zeros at the upper and lower sides of the passband can be created when L_c has a 180° phase difference with respect to M with similar magnitudes. This sign reversal is attributed to a destructive interference between two modes, therefore, resulting in the construction of transmission zeros at two frequencies.

The fundamental cross coupling technique is well explained in [85] by using multipath coupling diagrams to illustrate the relative phase shifts of multiple single paths. In [85], Brian adopted the S21 phase shift, $\Phi_{21,}$ of each lumped element in the equivalent circuits of a resonator and calculated the total phase shift at the input (or output) of the resonator to predict the behavior of transmission zeros. Since transmission zeros appear away from the passband, the off-resonance behavior of each lumped component is of concern. Based on Brian's theory, the equivalent circuit for a dual-mode cavity filter can be represented by a multipath diagram as described in Fig. 5.17(b). The shunt capacitor/inductor pairs of the equivalent circuit have been replaced by the black circles, and M

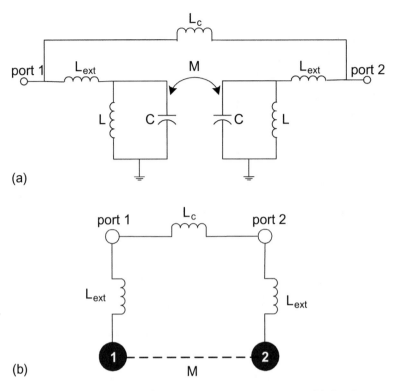

FIGURE 5.17: Equivalent-circuit model (a) and multicoupling diagram (b) for the quasi-elliptic dual-mode cavity filter.

represents the mutual electric coupling between two modes. The phase shift of each lumped element is used to calculate the total phase shift at the input or output of the filter for the different signal paths. In the case of the dual-mode single cavity filter, there are two possible signal paths (1) path 1: port 1-1-2-port 2 and (2) path 2: port 1-port 2. Both paths share the common input (port 1) and output (port 2). The total phase shifts for two signal paths in the dual-mode cavity are summarized in Table 5.3. The total phase shift for path 1 is $-90°$ both below and above resonance. The total phase shift for path 2 only accounts for the cross-magnetic coupling L_c between port 1 and port 2, hence being $+90°$. Therefore, two paths are out of phase both below and above resonance, meaning that destructive interferences creating transmission zeros occur both below and above the passband.

The locations of the upper and lower stop-band transmission zeros for the filter can be controlled by adjusting the values of M and L_c through varying the centerline offset, C_o, and distance, D_s, between the I/O external slots, respectively. The simulated responses of a dual mode filter as a function of the parameter C_o are shown in Fig. 5.18. When the feeding structure is placed at the center of the cavity ($C_o = 0$ mm), only the TE_{102} mode is excited, producing no transmission zeros. As C_o increases, the level of internal electric coupling, M, influences the upper transmission zeros

TABLE 5.3: Total phase shifts for two different paths in the dual-mode cavity filter.

PATHS	BELOW RESONANCE	ABOVE RESONANCE
Port 1-1-2-port 2	$-90° + 90° + 90° + 90° - 90°$	$-90° - 90° + 90° - 90° - 90°$
	$= +90°$	$= -270°$
Port 1-port 2	$-90°$	$-90°$
Result	Out of phase	Out of phase

more than the lower transmission zeros because of the asymmetrical effect of M upon the upper and lower poles [67]. The centerline offset, C_o, affects the performance of the 3-dB bandwidth and center frequency as well. It is observed that the maximum 3-dB bandwidth is obtained at the offset of 0.2 mm with the maximum coupling between dual modes.

Further increase of the offset results in a narrower bandwidth because the level of coupling for TE_{102} and TE_{201} changes. The downward shifting of the center frequency could be caused by the difference between the mean frequency $((f_o + f_e)/2)$ and the original resonant frequency of the cavity resonator. Also, external coupling can be attributed to the center frequency shift because of additional parasitic reactance from the feeding structures.

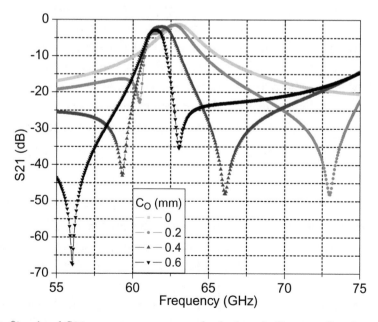

FIGURE 5.18: Simulated S21 parameter response of a dual mode filter as a function of the centerline offset C_o of the feeding structures.

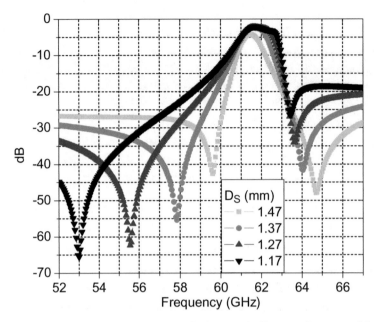

FIGURE 5.19: Simulated S21 parameter response of a dual mode filter as a function of the source-to-load distance D_s.

The transmission characteristic of the filter has also been investigated with respect to the values of L_c by varying the distance D_s between two external slots with a fixed centerline offset, C_o. Figure 5.19 displays the simulated response of a dual mode filter as a function of D_s with $C_o = 0.5$ mm. As L_c decreases by increasing D_s, the lower transmission zero shifts away from the center frequency while the higher transmission zero moves toward to the center frequency. The cross coupling, L_c, causes the asymmetrical shift of both transmission zeros due to the same reason mentioned in the case of M, influencing the lower transmission zero more than the higher one. The equivalent-circuit models validate the coupling mechanisms through the design of a transmitter filter in the next subsection.

5.4.1.5 Quasi-elliptic Dual-Mode Cavity Filter. Two dual-mode cavity filters exhibiting a quasiel-liptical response are presented as the next step for a three-dimensional integrated V-band transceiver front-end modules. The frequency range of interest is divided into two channels where the lower channel is allocated for an Rx, and the higher channel allocated for a Tx. To suppress the interference between the two channels as much as possible, the upper stop-band transmission zero of the Rx channel is placed closer to the center frequency of the passband than the lower stop-band zero. In the case of a Tx filter, the lower zero is located closer to the center frequency of the passband than the upper zero.

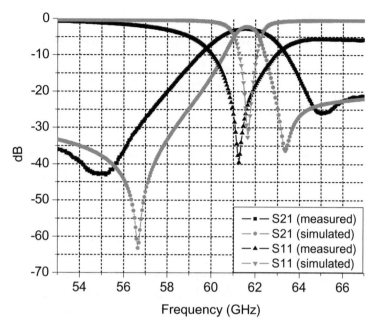

FIGURE 5.20: Measured and simulated S-parameters of the dual-mode cavity filter for an Rx channel.

First, a Rx filter was designed and validated with experimental data, as shown in Fig. 5.20. A line-reflect-reflect-match (LRRM) method [86] was employed for calibration of the measurements with 250μm pitch air coplanar probes. In the measurement, the reference planes were placed at the end of the probing pads, and the capacitance and inductance effects of the probing pads were de-embedded by use of "Wincal" software so that effects, such as those due to the CPW loading, become negligible. The filter exhibits an insertion loss of <2.76 dB, center frequency of 61.6 GHz, and 3-dB bandwidth of about 4.13% ($\approx$2.5 GHz). The upper and lower transmission zeros are observed to be within 3.4 GHz and 6.4 GHz away from the center frequency, respectively.

Then, a Tx filter using a dual-mode cavity resonator was designed for a center frequency of 63.4 GHz, fractional 3-dB bandwidth of 2%, insertion loss of <3 dB, and 25 dB rejection bandwidth on the lower side of the passband of <2 GHz. To obtain a center frequency of 63.4 GHz, the size of the via-based cavity was adjusted and determined to be 2.04 × 2.06 × 0.106 ($L \times W \times H$ in Fig. 5.13) mm^3. The corresponding lumped-element values in the equivalent-circuit model [Fig. 5.17(a)] of a Tx filter were evaluated, and their values were $L_{ext} = 0.074$ nH, $L = 0.0046$ nH, $C = 1.36$ pF, $M = 0.032$ pF and $L_c = 0.73$ nH. Figure 5.21(a) shows the ideal response from the circuit model, exhibiting two transmission zeros at 61.6 and 68.7 GHz. The measured insertion loss and reflection losses of the fabricated filter are compared to the full-wave simulation results in Fig. 5.21(b). The fabricated Tx filter exhibits an insertion loss of 2.43 dB, which is slightly higher than the simulated loss (2.0 dB). The main source of this discrepancy might be caused by the skin and edge effects

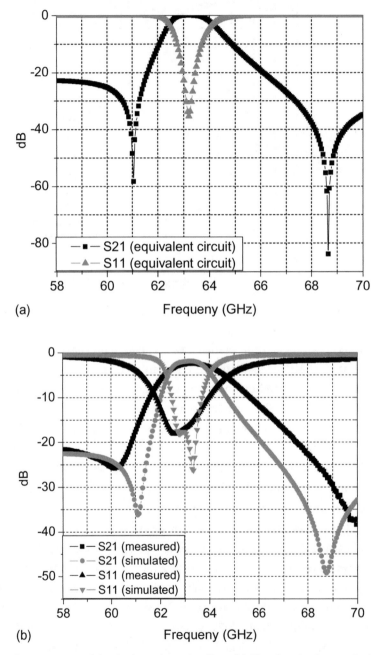

FIGURE 5.21: S-parameters of the dual-mode cavity filter. (a) Simulated using equivalent-circuit model in Fig. 17(a). (b) Measured and simulated for a Tx channel.

TABLE 5.4: Design parameters of quasielliptic dual-mode cavity filters.

DESIGN PARAMETERS	RX FILTER (mm)	TX FILTER (mm)
Cavity length (L)	2.075	2.04
Cavity width (W)	2.105	2.06
Cavity height (H)	0.106	0.106
External slot length (E_L)	0.360	0.360
External slot width (E_W)	0.572	0.572
Centerline offset (C_o)	0.5675	0.35
Distance between external slots (D_s)	1.37	1.355

of the metal traces since the simulations assume a perfect definition of metal strips with finite thickness.

The center frequency was measured to be 63.4 GHz, which is in good agreement with the simulated result. The upper and lower transmission zeros were observed to be within 6.5 and 3.2 GHz away from the center frequency, respectively. Those can be compared to the simulated values that exhibit the upper and lower transmission zeros within less than 5.3 and 2.3 GHz away from the center frequency. The discrepancy of the zero positions between the measurement and the simulation can be attributed to the fabrication tolerance. Also, the misalignment between the substrate layers in the LTCC process might cause an undesired offset of the feeding structure position. This could be another significant reason for the transmission zero shift. The fabrication tolerances also result in the bandwidth differences. The filter exhibits a 3-dB measured bandwidth of 4.02% (~2.5 GHz) compared to the simulated one of 2% (~1.3 GHz). All of the final layout dimensions optimized using HFSS are summarized in Table 5.4.

5.4.2   Multipole Dual-Mode Cavity Filters

In order to provide the additional design guidelines for generic multipole cavity filters, the authors proceed with a vertically stacked arrangement of two dual-mode cavities. The presynthesized dual-mode cavities are stacked with a coupling slot in order to demonstrate the feasibility of realizing a multipole filter by using the dual-mode cavity filters investigated in Section 5.4.1. Two well-known types of slots (rectangular and cross-shaped) are considered as the intercoupling structure in this study. In the past, mode matching methods [70] and scattering matrix approaches [76] have been used to analyze the modal characterization of intercoupling discontinuities hence will not be covered here.

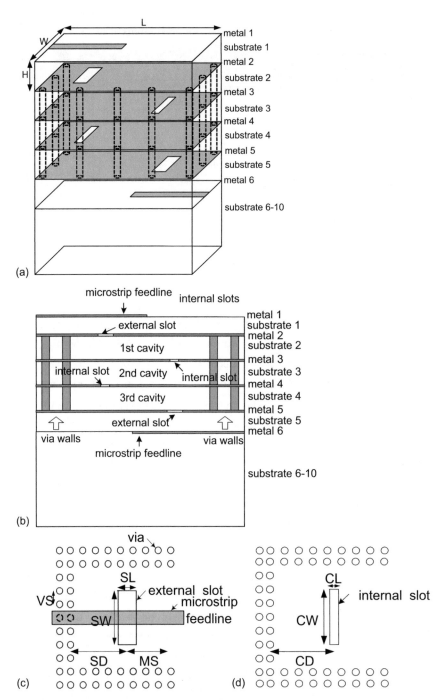

FIGURE 5.22: 3D overview (a) and top view (b) of a vertically stacked multipole dual-mode cavity filter. (c) Intercoupling rectangular slot (d) Intercoupling cross slot.

The 3D overview, top view, intercoupling rectangular slot, and intercoupling cross slot of the proposed cavity filter are illustrated in Fig. 5.22(a)–(d). The top five substrate layers [microstrip line: S1, cavity 1: S2–S3, cavity 2: S4–S5 in Fig. 5.22(a)] are occupied by the filter. Microstrip lines have been employed as the I/O feeding structure on the top metal layer, M1, and excite the first dual-mode cavity through the rectangular slots on the top ground plane, M2, of the cavity 1. Two identical dual-mode cavity resonators [cavity 1 and cavity 2 in Fig. 5.22(a)] are vertically stacked and coupled through an intercoupling slot to achieve the desired frequency response with high selectivity as well as a high-level of compactness.

5.4.2.1 Quasielliptic Filter with a Rectangular Slot. The multipath diagram of a vertically stacked dual-mode filter with a rectangular slot is illustrated in Fig. 5.23. The black circles denoted by 1 and 2 are the degenerate resonant modes in the top dual-mode cavity while the one denoted by 3 represents the excited resonant mode in the bottom cavity. The coupling, M_{12}, is realized through the electrical coupling and is controlled by the offsets of the I/O feeding structures. Also, the intercouplings, M_{13} and M_{32}, are determined by the size and position of the intercoupling slots and dominated by the magnetic coupling. It is worth noting that M_{13} is different from M_{32} since the magnitude of the magnetic dipole moment of each mode in a coupling slot is different to each other due to the nature of a rectangular slot. Since the rectangular slot is parallel to the horizontal direction, the modes polarized to the horizontal direction are more strongly coupled through the slot than the modes that are polarized in the vertical direction. However, by adjusting the offset, we attempted to obtain the appropriate coupling level of M_{13} and M_{32} to realize the desired filter response. L_c (the magnetic coupling parameter) is used to implement the cross coupling between port 1 and port 2. The phase shifts for three possible signal paths are summarized in Table 5.5. The filter with three modes can

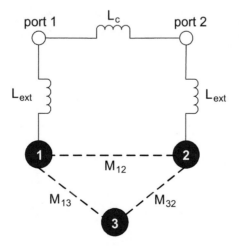

FIGURE 5.23: Multicoupling diagram for the vertically stacked multipole dual-mode cavity filter with a rectangular slot for intercoupling between two cavities.

TABLE 5.5: Total phase shifts for three different signal paths in the vertically stacked dual-mode cavity filter with a rectangular slot.

PATHS	BELOW RESONANCE	ABOVE RESONANCE
Port 1-1-2-port 2	$-90° + 90° + 90° + 90° - 90°$ $= +90°$	$-90° - 90° + 90° - 90° - 90°$ $= -270°$
Port 1-port 2	$-90°$	$-90°$
Result	Out of phase	Out of phase
1-3-2	$-90° + 90° - 90° = -90°$	$-90° - 90° - 90° = -270°$
1-2	$+90°$	$+90°$
Result	Out of phase	In phase

generate two transmission zeros below resonance and an additional zero above resonance.[move this sentence to the previous paragraph!]

The three-pole quasi-elliptic filters were designed to meet the following specifications: (1) center frequency: 66 GHz, (2) 3-dB fractional bandwidth: ~2.6%, (3) insertion loss: <3 dB, and (4) 15 dB rejection bandwidth using triple transmission zeros (two on the lower side and one on the

TABLE 5.6: Design parameters of multipole dual-mode cavity filters with two types of inter-coupling slots.

DESIGN PARAMETERS	RECTANGULAR (mm)	CROSS (mm)
Cavity length (L)	2.04	2.06
Cavity width (W)	1.92	2.06
Each cavity height (H)	0.106	0.106
External slot length (E_L)	0.440	0.470
External slot width (E_W)	0.582	0.472
Centerline offset (C_o)	0.245	0.356
Internal slot length (I_L)	0.642	0.412
Internal slot width (I_W)	0.168	0.145
Vertical slot offset (V)	0.325	0.6075
Horizontal slot offset (R)	0.065	0
Distance between external slots (D_s)	1.29	1.26

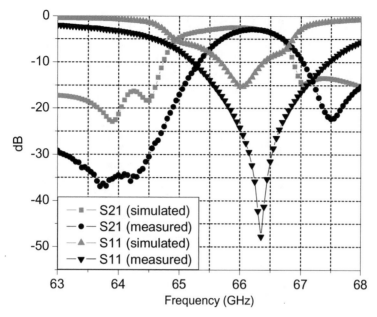

FIGURE 5.24: Measured and simulated S-parameters of the quasielliptic dual-mode cavity filter with a rectangular slot for inter coupling between cavities.

upper side): <3 GHz. A study of the dual-mode coupling in each cavity on the basis of the initial determination of the cavity size resonating at a desired center frequency (66 GHz) is performed first. Then, the final configuration of the three-pole dual-band filter can be obtained through the optimization of the intercoupling slot size and offsets via simulation.

All the design parameters for the filters are summarized in Table 5.6. Figure 5.24 shows the measured performance of the designed filters with a rectangular slot along with a comparison to the simulated results. It can be observed that the measured results in the case of a rectangular slot produce a center frequency of 66.2 GHz with the bandwidth of 1.2 GHz (~1.81%), and the minimum insertion loss in the passband around 2.9 dB. The simulation showed a minimum insertion loss of 2.5 dB with a slightly wider 3-dB bandwidth of 1.7 GHz (~2.58%) around the center frequency of 65.8 GHz. The center frequency shift is caused by XY shrinkage of ±3%. The two measured transmission zeros with a rejection better than 34 dB and 37 dB are observed within <1.55 GHz and <2.1 GHz, respectively, away from the center frequency at the lower band than the passband. One transmission zero is observed within <1.7 GHz at the higher band than the passband. The discrepancy of the zero positions and rejection levels between the measurement and the simulation can be attributed to the fabrication tolerances as explained in Section 5.4.1.5. Still, it can be observed that the behavior of transmission zeros shows a good correlation of measurements and simulations.

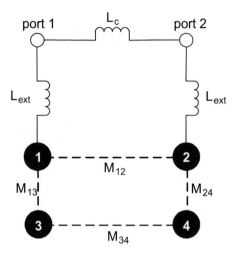

FIGURE 5.25: Multicoupling diagram for the vertically stacked multipole dual-mode cavity filter with a rectangular slot for intercoupling between two cavities.

This type of filter can be used to generate the sharp skirt at the lower side to reject local oscillator and image signals as well the extra transmission zero in the high skirt that can be utilized to suppress the harmonic frequencies according to the desired design specifications.

5.4.2.2 Quasi-elliptic Filter with a Cross Slot. The cross slot is applied as an alternative intercoupling slot between the two vertically stacked cavities. The multipath diagram for the filter and the phase shifts for the possible signal paths are described in Fig. 5.25 and Table 5.7. Each cavity supports two

TABLE 5.7: Total phase shifts for three different signal paths in the vertically stacked dual-mode cavity filter with a cross slot.

PATHS	BELOW RESONANCE	ABOVE RESONANCE
Port 1-1-2-port 2	$-90° + 90° + 90° + 90° - 90°$	$-90° - 90° + 90° - 90° - 90°$
	$= +90°$	$= -270°$
Port 1-port 2	$-90°$	$-90°$
Result	Out of phase	Out of phase
1-3-4-2	$-90° + 90° + 90° + 90 - 90$	$-90° - 90° + 90° - 90 - 90$
	$= +90°$	$= -270°$
1-2	$+90°$	$+90°$
Result	In phase	In phase

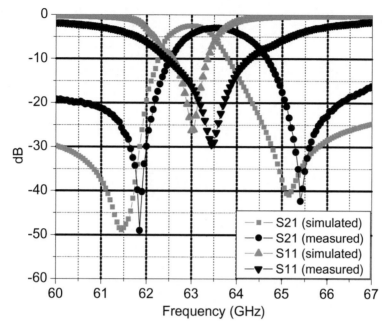

FIGURE 5.26: Measured and simulated S-parameters of the quasi-elliptic dual-mode cavity filter with a rectangular slot for inter coupling between cavities.

orthogonal dual modes (1 and 2 in the top cavity, 3 and 4 in the bottom cavity) since the cross-slot structure excites both degenerate modes in the bottom cavity by allowing the coupling between the modes that have the same polarizations. The coupling level can be adjusted by varying the size and position of the cross slots. The couplings of M_{12} and M_{34} are realized by electrical coupling while the inter couplings of M_{13} and M_{24} are realized by magnetic coupling. The total phase shifts of the four signal paths of the proposed structure prove that they generate one zero above resonance and one below resonance.

The quasielliptic filters were designed for a sharp selectivity. The simulation achieved the following specifications: (1) Center frequency: 63 GHz, (2) 3-dB fractional bandwidth: ~2%, (3) Insertion loss: <3 dB, and (4) 40 dB rejection bandwidth using two transmission zeros (one on the lower side and one on the upper side): <4 GHz. The filter was fabricated using LTCC substrate layers. Figure 5.26 shows the measured results compared to those of the simulated design. The fabricated filter exhibits a center frequency of 63.5 GHz, an insertion loss of approximately 2.97 dB, a 3-dB bandwidth of approximately 1.55 GHz (~2.4%), and >40 dB rejection bandwidth of 3.55 GHz.

CHAPTER 6

Three-Dimensional Antenna Architectures

6.1 SOFT-SURFACE STRUCTURES FOR IMPROVED-EFFICIENCY PATCH ANTENNAS

The radiation performance of patch antennas on large-size substrate can be significantly degraded by the diffraction of surface waves at the edge of the substrate. Most modern techniques for the surface-wave suppression are related to periodic structures, such as photonic band-gap (PBG) or electromagnetic band-gap (EBG) geometries [87–89]. However, those techniques require a considerable area to form a complete band-gap structure. In addition, it is usually difficult for most printed-circuit technologies to realize such a perforated structure. In this chapter, we present the novel concept of the "soft surface" to improve the radiation pattern of patch antennas [90]. A single square ring of shorted quarter-wavelength metal strips is employed to form a soft surface and to surround the patch antenna for the suppression of outward propagating surface waves, thus alleviating the diffraction at the edge of the substrate. Since only a single ring of metal strips is involved, the formed "soft surface" structure is compact and easily integrable with three-dimensional (3D) modules.

6.1.1 Investigation of an Ideal Compact Soft Surface Structure

For the sake of simplicity, we consider a probe-fed square patch antenna operating at 15 GHz on a square grounded substrate with thickness H ($\sim 0.025\lambda_0$, λ_0 is the free-space wavelength) and a dielectric constant ε_r (~ 5.4). The patch antenna is surrounded by the ideal compact soft surface that consists of a square ring of metal strip, that are short-circuited to the ground plane by a metal wall along the outer edge of the ring, as shown in Fig. 6.1.

The inner length of every side of the soft surface ring (denoted by L_s) was found to be approximately one wavelength plus L_p. The substrate's size is assumed to be $L \times L$ ($2\lambda_0 \times 2\lambda_0$), much larger than the size ($L_p \times L_p < 0.5\lambda_g \times 0.5\lambda_g$) of the square patch. The width of the metal strip (W_s) is approximately equal to one quarter of the guided wavelength. The mechanism for the radiation pattern improvement achieved by the introduction of a compact soft surface structure can be understood by considering two factors. First the quarter-wave shorted metal strip serves as an open circuit for the TM_{10} mode (the fundamental operating mode for a patch antenna). Therefore, it is difficult for the surface current on the inner edge of the soft surface ring to flow outward

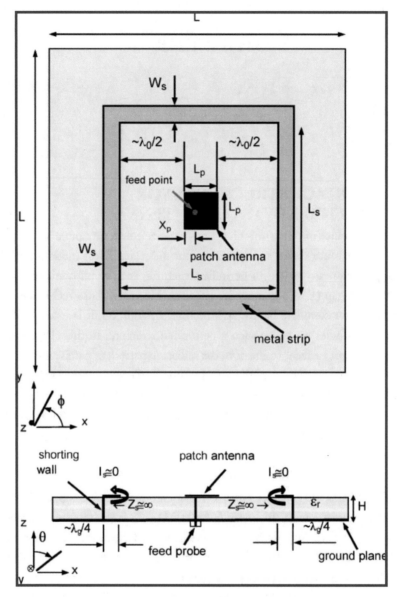

FIGURE 6.1: Patch antenna surrounded by an ideal compact soft surface structure consisting of a ring of metal strip and a ring of shorting wall (I_s, surface current on the top surface of the soft surface ring, Z_s, impedance looking into the shorted metal strip).

(also see Fig. 6.2). As a result, the surface waves can be considerably suppressed outside the soft surface ring, hence reducing the undesirable diffraction at the edge of the grounded substrate.

This explanation can be confirmed by checking the field distribution in the substrate. Figure 6.2 shows the electric field distributions on the top surface of the substrate for the patch antennas with

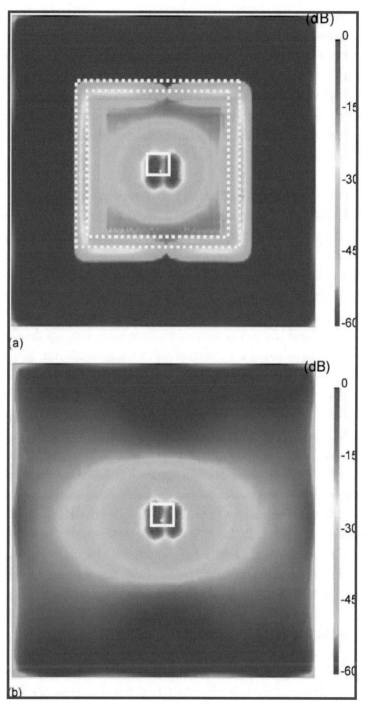

FIGURE 6.2: Simulated electric field distributions on the top surface of the substrate for the patch antennas with (a) and without (b) the soft surface ($\varepsilon_r = 5.4$).

and without the soft surface. We can see that the electric field is indeed contained inside the soft surface ring. It is estimated that the field magnitude outside the ring is approximately 5 dB lower than that without the soft surface. The second factor contributing to the radiation pattern improvement is the fringing field along the inner edge of the soft surface ring. This fringing field along with the fringing field at the radiating edges of the patch antenna forms an antenna array in the E-plane. The formed array acts as a broadside array with minimum radiation in the x-y plane when the distance between the inner edge of the soft surface ring and its nearby radiating edge of the patch is roughly half a wavelength in free space.

6.1.2   Implementation of the Soft-Surface Structure in LTCC

To demonstrate the feasibility of the multilayer LTCC technology on the implementation of the soft surface, we first simulated a benchmarking prototype that was constructed replacing the shorting wall with a ring of vias. The utilized low temperature cofired ceramic (LTCC) material had a dielectric constant of 5.4. The whole module consists of a total of 11 LTCC layers (layer thickness $= 100\,\mu$m) and 12 metal layers (layer thickness $= 10\,\mu$m). The diameter of each via specified by the fabrication process was $100\,\mu$m, and the distance between the centers of two adjacent vias was $500\,\mu$m. To support the vias, a metal pad is required on each metal layer; to simplify the simulation, all pads on each metal layer are connected by a metal strip with a width of $600\,\mu$m. Simulation shows that the width of the pad metal strips has little effect on the performance of the soft surface structure as long as it is less than the width of the metal strips for the soft surface ring (W_s). The size of the LTCC board was 30 mm × 30 mm. The operating frequency was set within the K_u-band (the design frequency $f_0 = 16.5$ GHz).

The optimized values for L_s and W_s were, respectively, 22.2 mm and 1.4 mm, which led to a total via number of 200 (51 vias on each side of the square ring). Including the width ($300\,\mu$m) of the pad metal strip, the total metal strip width for the soft surface ring was found to be 1.7 mm. Since the substrate was electrically thick at $f_0 = 16.5$ GHz ($>0.1\lambda_g$), a stacked configuration was adopted for the patch antenna to improve its input impedance performance. By adjusting the distance between the stacked square patches, a broadband characteristic for the return loss can be achieved [91]. For the present case, the upper and lower patches (with the same size 3.4 mm × 3.4 mm) were respectively printed on the first LTCC layer and the seventh layer from the top, leaving a distance between the two patches of 6 LTCC layers. The lower patch was connected by a via hole to a 50-Ω microstrip feed line that was on the bottom surface of the LTCC substrate. The ground plane was embedded between the second and third LTCC layers from the bottom. The inner conductor of an SMA (semi-miniaturized type-A) connector was connected to the microstrip feed line while its outer conductor was soldered on the bottom of the LTCC board to a pair of pads that were shorted to the ground through via metallization. It has to be noted that the microstrip feed line was printed on the bottom of the LTCC substrate to avoid its interference with the soft surface ring and to

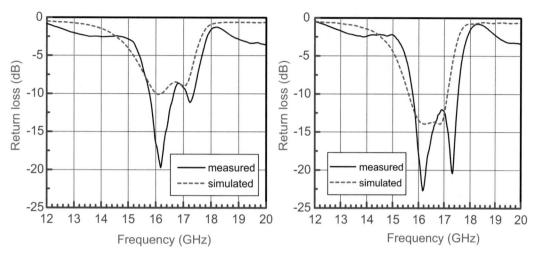

FIGURE 6.3: Comparison of return loss between simulated and measured results for the stacked-patch antennas with (a) and without (b) the soft surface implemented on LTCC technology.

alleviate the contribution of its spurious radiation to the radiation pattern at broadside. An identical stacked-patch antenna on the LTCC substrate without the soft surface ring was also built for comparison.

The simulated and measured results for the return loss shown in Fig. 6.3 show good agreement. As the impedance performance of the stacked-patch antenna is dominated by the coupling between the lower and upper patches, the return loss for the stacked-patch antenna seems more sensitive to the soft surface structure than that for the previous thinner single patch antenna. The measured return loss is close to $-10\,dB$ over the frequency range 15.8–17.4 GHz (about 9% in bandwidth). The slight discrepancy between the measured and simulated results is mainly due to the fabrication issues (such as the variation of dielectric constant or/and the deviation of via positions) and the effect of the transition between the microstrip line and the SMA (SubMiniature version A) connector.

It is also noted that there is a frequency shift of about 0.3 GHz (about 1.5% up). This is probably caused by the LTCC material that may have a real dielectric constant slightly lower than the over estimated design value. It is noted that it is normal for practical dielectric substrates to have a dielectric constant within a ±2% deviation.

The radiation patterns measured in the E- and H-planes show a good agreement with the simulation with the simulated results in Fig. 6.4 for the frequency of 17 GHz where the maximum gain of the patch antenna with the soft surface was observed. It is confirmed that the radiation at broadside is enhanced and the backside level is reduced. Also the beam width in the E-plane is significantly reduced by the soft surface, realizing a more directive radiation performance. It is noted

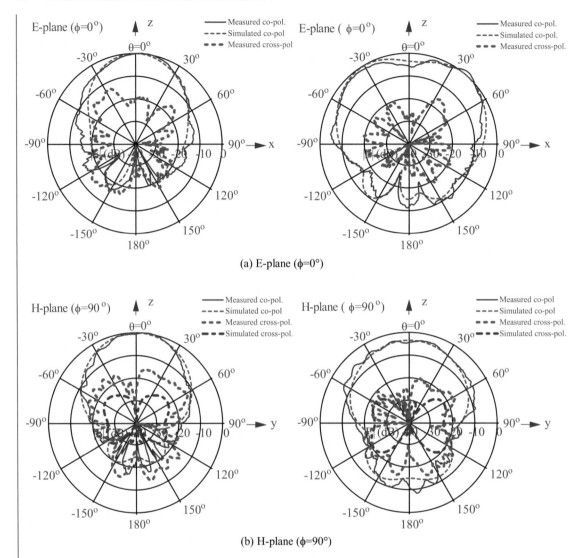

FIGURE 6.4: Comparison between simulated and measured radiation patterns for the stacked-patch antennas with (*left*) and without (*right*) the soft surface implemented on LTCC technology ($f_0 = 17$ GHz). (a) E-plane ($\phi = 0°$). (b) H-plane ($\phi = 90°$).

that the measured cross-polarized component has a higher level and more ripples than the simulation result. This is because the simulated radiation patterns were plotted in two ideal principal planes, i.e., $\phi = 0°$ and $\phi = 90°$ planes. The simulations demonstrated that the maximum cross-polarization may happen in the plane $\phi = 45°$ or $\phi = 135°$. During measurement, a slight deviation from the ideal planes can cause a considerable variation for the cross-polarized component since the spatial variation of the cross-polarization is quick and irregular.

Also, a slight polarization mismatch or/and some objects near the antenna (such as the connector or/and the connection cable) may considerably contribute to the high cross-polarization. In addition, the maximum gain measured for the patch with the soft surface is near 9 dBi, about 3 dB higher than the maximum gain and 7 dB higher than the gain at broadside for the antenna without the soft surface.

6.2   HIGH-GAIN PATCH ANTENNA USING A COMBINATION OF A SOFT-SURFACE STRUCTURE AND A STACKED CAVITY

The advanced technique of the artificial soft surface consisting of a single square ring of metal strip shorted to the ground demonstrated the advantages of compact size and excellent improvement in the radiation pattern of patch antennas in section 6.1. In this section, we further improve this technique by adding a cavity-based feeding structure on the bottom LTCC layers [substrate 4 and 5 in Fig. 6.5(c)] of an integrated module to increase the gain even more and to reduce future backside radiation. The maximum gain for the patch antenna with the soft surface and the stacked cavity is approximately 7.6 dBi that is 2.4 dB higher than 5.2 dBi for the "soft-enhanced" antenna without the backing cavity.

6.2.1   Antenna Structure Using a Soft-Surface and Stacked Cavity

The 3D overview, top view and cross-sectional view of the topology chosen for the micostrip antenna using a soft-surface and a vertically stacked cavity are shown in Fig. 6.5(a), (b) and (c), respectively. The antenna is implemented into five LTCC substrate layers (layer thickness $= 117\,\mu$m) and six metal layers (layer thickness $= 9\,\mu$m). The utilized LTCC is a novel composite material of high dielectric constant ($\varepsilon_r\sim7.3$) in the middle layer (substrate 3 in Fig. 6.5(c)) and slightly lower dielectric constant ($\varepsilon_r\sim7.0$) in the rest of the layers [substrate 1–2 and 4–5 in Fig. 6.5(c)]. A 50 Ω stripline is utilized to excite the microstrip patch antenna (metal 1) through the coupling aperture etched on the top metal layer (metal 4) of the cavity as shown in Fig. 6.5(c). In order to realize the magnetic coupling by maximizing magnetic currents, the slot line is terminated with a $\lambda_g/4$ open stub beyond the slot.

The probe feeding mechanism could not be used as a feeding structure because the size of the patch at the operating frequency of 61.5 GHz is too small to be connected to a probe via according to the LTCC design rules. The patch antenna is surrounded by a soft surface structure consisting of a square ring of metal strips that are short-circuited to the ground plane [metal *4* in Fig. 6.5(c)] for the suppression of outward propagating surface waves. Then, the cavity [Fig. 6.5(c)], that is realized utilizing the vertically extended via fences of the "soft surface" as its sidewalls, is stacked right underneath the antenna substrate layers [substrates 4 and 5 in Fig. 6.5(c)] to further improve the gain and to reduce backside radiation. The operating frequency is chosen to be 61.5 GHz; the optimized size ($P_L \times P_W$) of patch is 0.54×0.88 mm^2 with the rectangular coupling slot ($S_L \times S_W = 0.36 \times 0.74$ mm^2). The size ($L \times L$) of the square ring and the cavity is optimized to be 2.6×2.6 mm^2 to achieve the

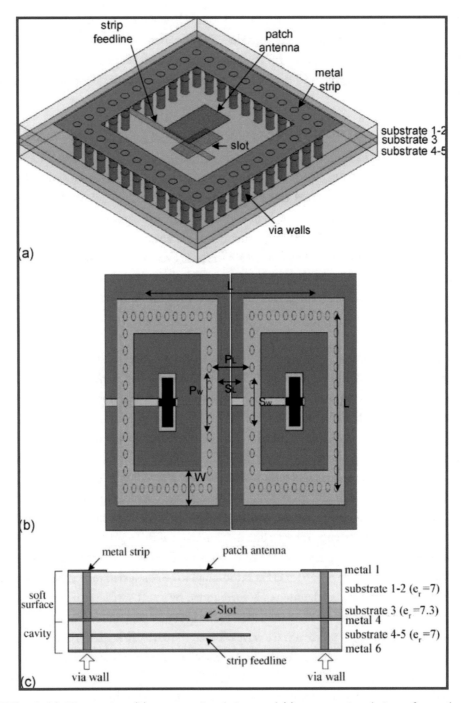

FIGURE 6.5: (a) 3D overview, (b) cross-sectional view, and (c) cross-sectional view of a patch antenna with the soft surface and stacked cavity.

maximum gain. The width of metal strip (W) is found to be 0.52 mm to serve as an open circuit for the TM_{10} mode of the antenna. The ground planes are implemented on metals 4 and 6.

We achieved the significant miniaturization on the ground planes because their size excluding the feeding lines is the same as that of the soft surface ($\approx 3.12 \times 3.12$ mm^2). In addition, the underlying cavity is used both as a dual-mode filter to separate the TM_{10} mode whose phase and amplitude contain the information transmitted through short-range indoor wireless personal area network (WPAN) and as a reflector to improve the gain.

6.2.2   Simulation and Measurement Results

The simulated (HFSS) and the measured results for the return loss are shown in Fig. 6.6. The measured return loss is close to -10 dB over the frequency range 58.2–62.3 GHz (about 6.6% in bandwidth). The slight discrepancy between the measured and simulated results is mainly due to the fabrication issues, such as the variation of dielectric constant or/and the deviation of via positions. From our investigation on the impedance performance, it is noted that the soft-surface structure vertically stacked by the cavity does not affect significantly on the bandwidth of the patch.

We compared the gains among the patch antennas with the soft surface and the stacked cavity, with the soft surface only, and without the soft surface. The simulated gains at broadside (i.e., the z-direction) are shown in Fig. 6.7. The simulated gain was obtained from the numerically calculated directivity in the z-direction and the simulated radiation efficiency, which is defined as the radiated

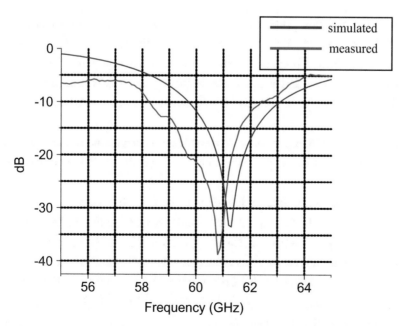

FIGURE 6.6: Comparison of return loss between simulated and measured results for a patch antenna with the soft surface and the stacked cavity implemented on LTCC technology.

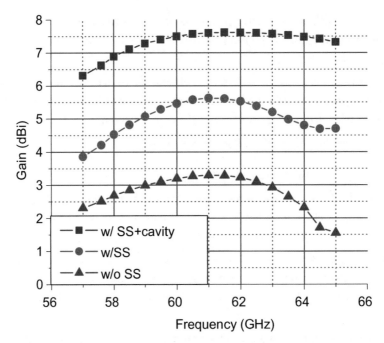

FIGURE 6.7: Comparison of simulated and measured gains at broadside between the stacked-patch antennas with and without the soft surface (SS) implemented in LTCC technology.

power divided by the radiation power plus the ohmic loss from the substrate and metal structures ($\tan \delta = 0.0024$ and $\sigma = 5.8 \times 10^7$ S/m were assumed for the Copper metallization). In Fig. 6.7, we can see that the simulated broadside gain of the patch antenna with the soft surface and the stacked cavity is more than 7.6 dBi at the center frequency, about 2.0 dB improvement as compared to one with the soft surface only and 4.3 dB improvement as compared to one without the soft surface.

More gain enhancement is possible with the thicker substrate since the thicker substrate excites stronger surface waves while the soft surface blocks and transforms the excited surface waves into space waves.

The radiation patterns simulated in E and H planes of patch antennas with the soft surface only and with the soft surface/stacked cavity are shown and compared in Fig. 6.8(a) and (b), respectively. The radiation patterns compared here are for a frequency of 61.4 GHz where the maximum gain of the patch antenna with the soft surface was observed. It is confirmed that the radiation at broadside is enhanced by 2.4 dB and the backside level is significantly reduced by 5.1 dB by stacking the cavity to the patch antenna with the soft surface. Also the beam width in the E-plane is reduced from 74° to 68° with the addition of the staked cavity.

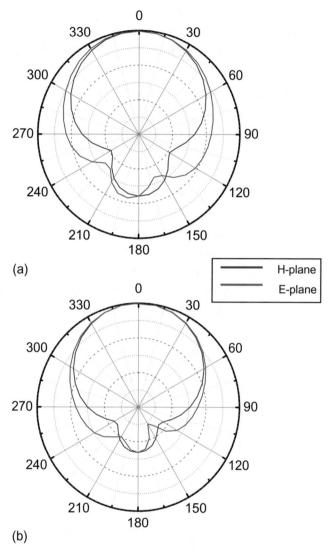

FIGURE 6.8: Radiation characteristics at 61.5 GHz of patch antennas (a) with the soft surface and (b) with the soft surface and the stacked cavity.

6.3   DUAL-POLARIZED CROSS-SHAPED MICROSTRIP ANTENNA

The next presented antenna for an easy integration with 3D modules is a cross-shaped antenna, that was designed for the transmission and reception of signals that cover two bands between 59–64 GHz. The first band (channel 1) covers 59–61.25 GHz, while the second band (channel 2) covers 61.75–64 GHz. Its structure is dual-polarized for the purpose of doubling the data output rate transmitted and received by the antenna. The cross-shaped geometry was utilized to decrease the cross-polarization that contributes to unwanted side lobes in the radiation pattern [92].

6.3.1   Cross-Shaped Antenna Structure

The antenna, shown in Fig. 6.9, was excited by proximity-coupling and had a total thickness of 12 metal layers and 11 substrate layers (each layer was 100 μm thick). Proximity-coupling is a particular method for feeding patch antennas where the feedline is placed on a layer between the antenna and the ground plane. When the feedline is excited, the fringing fields at the end of the line strongly couple to the patch by electromagnetic coupling. This configuration is a non-contact, non-coplanar method of feeding a patch antenna, that allows different polarization reception of signals that exhibits improved cross-channel isolation in comparison to a traditional coplanar microstrip feed.

There were two substrate layers separating the patch and the feedline, and two substrate layers separating the feedline and the ground layer. The remaining seven-substrate layers were used for embedding the radio frequency (RF) circuitry beneath the antenna; that includes the filter, integrated passives and other components. The size of the structure was $8 \times 7\,\text{mm}^2$. A right angle bend in the feedline of channel 2 is present for the purpose of simplifying the scattering parameter measurements on the network analyzer.

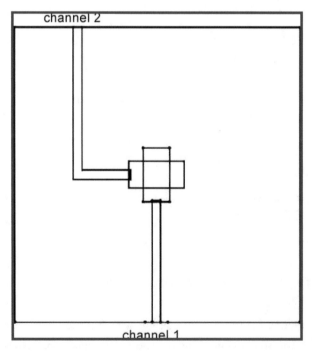

FIGURE 6.9: Cross-shaped antenna structure in LTCC.

6.3.2 Simulation and Measurement Results

Figure 6.10(a) shows the simulated scattering parameters versus frequency for this design. The targeted frequency of operation was around 60.13 GHz for channel 1 (S11) and 62.87 GHz for channel 2 (S22). The simulated return loss for channel 1 was close to -28 dB at $f_r = 60.28$ GHz, while for channel 2, the return loss was ~ -26 dB at $f_r = 62.86$ GHz. The simulated frequency for channel 1 was optimized in order to cover the desired band based on the antenna structure. Channel 2 has a slightly greater bandwidth (3.49%) than that of channel 1 (3.15%) primarily due to the right angle bend in the feedline that can cause small reflections to occur at neighboring frequencies near the resonance point of the lower band. The upper edge frequency (f_H) of the lower band is 61.21 GHz; while the lower edge frequency (f_L) of the higher band is 61.77 GHz.

Figure 6.10(b) shows the measured scattering parameters versus frequency for the design. The measured return loss for channel 1 (-20 dB at $f_r = 58.5$ GHz) is worse than that obtained through the simulation (-26 dB). Conversely, the -40 dB of measured return loss at $f_r = 64.1$ GHz obtained for channel 2 is significantly better than the simulated return loss of -28 dB. The diminished return loss of channel 1 is acceptable due to minor losses associated with measurement equipment (cables, connectors, etc.). The enhanced return loss of channel 2 could result from measurement inaccuracies or constructive interference of parasitic resonances at or around the TM_{10} resonance. The asymmetry in the feeding structure may account for this difference in the measured return loss. Frequency shifts for both channels are present in the measured return loss plots. Additionally, the bandwidths of the two channels are wider than those seen in simulations (5.64% for channel 1 and 8.26% for channel 2).

Small deviations in the dimensions of the fabricated design as well as measurement tolerances may have contributed to the frequency shifts, while the increased bandwidths may be attributed to radiation from the feedlines and other parasitic effects that resonate close to the TM_{10} mode producing an overall wider bandwidth. The upper edge frequency (f_H) of the lower band is 61 GHz, while the lower edge frequency (f_L) of the higher band is 62.3 GHz. The simulated cross coupling between channels 1 and 2 (Fig. 6.10) is below -22 dB for the required bands. On the other hand, the measured cross coupling between the channels is below -22 dB for the lower band and below -17 dB for the upper band. Due to the close proximity of the feeding line terminations of the channels, the cross coupling is hindered, but these values are satisfactory for this application.

6.4 SERIES-FED ANTENNA ARRAY

The last example presented in this section deals with a compact antenna array, that could potentially find application in numerous MIMO systems or point-to-point/point-to-multipoint multimedia (e.g. wireless HDTV). Specifically a series fed 1×4 linear antenna array of four microstrip patches [93], covering the 59–64 GHz band, which has been allocated world wide for dense wireless local communications [94], has been designed on LTCC substrate.

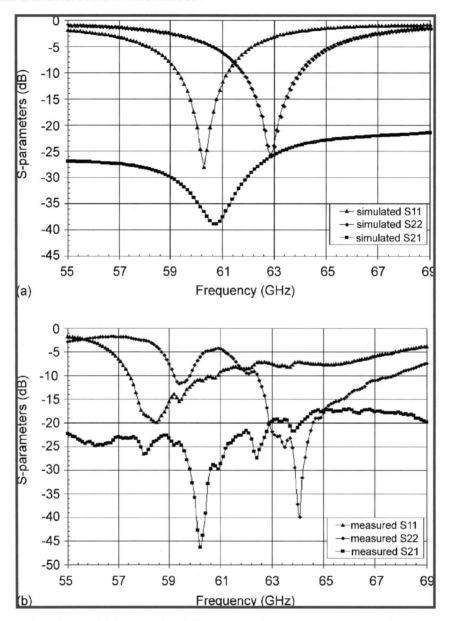

FIGURE 6.10: (a) Simulated and (b) measured S-parameter data versus frequency.

6.4.1   Antenna Array Structure

The top and cross-sectional views of a series-fed 1×4 linear antenna array are illustrated in Fig. 6.11(a) and (b), accordingly. The proposed antenna employs a series feed instead of a corporate feed because of its easy-to-design feeding network and low level of radiation from the feed line [93]. The matching between neighboring elements is achieved by controlling the width

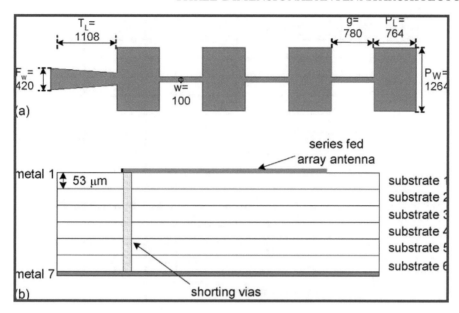

FIGURE 6.11: (a) Top view and (b) cross-sectional view of a series fed 1×4 linear array of four microstrip patches. All dimensions indicated in (a) are in micrometers.

(P_W in Fig. 6.11(a)) of the patch elements. The antenna was screen-printed on the top metal layer [metal 1 in Fig. 6.11(b)], and uses six substrate layers to provide the required broadband matching property and high gain. The targeted operation frequency was 61.5 GHz. First, the single patch resonator ($0.378\lambda_g \times 0.627\lambda_g$) resonating at 61.5 GHz is designed. The width-to-line ratio of the patch is determined to obtain the impedance matching and the desired resonant frequency. In our case, identical four patch resonators are linearly cascaded using thin microstrip lines [$w = 0.100$ mm in Fig. 6.11(a)] to maximize the performance at the center frequency of 61.5 GHz. The distance [g in Fig. 6.11(a)] between patch elements is the critical design parameter to achieve equal amplitude and cophase (equal phase) excitation and control the tilt of the maximum beam direction. It was optimized to be 0.780 mm ($\sim0.387\lambda_g$) for $0°$ tilted fan beam antenna. The physical length of the tapered feeding line was determined to be 1.108 mm (T_L in Fig. 6.11(a)).

6.4.2   Simulation and Measurement Results

Figure 6.12 demonstrates the very good correlation between the measured and simulated return loss (S11) versus frequency for this design. The measured 10-dB BW is 55.4–66.8 GHz ($\sim18.5\%$) compared to the simulated that is 54–68.4 GHz ($\sim23.4\%$). The narrower BW might be due to the band limiting effect from the coplanar waveguide (CPW) measurement pad (0.344×1.344 mm^2).

Figure 6.13 presents E-plane and H-plane radiation patterns at the center frequency of 61.5 GHz. We can easily observe the $0°$ beam tilt from the radiation characteristics. The maximum gain of this antenna is 12.6 dBi.

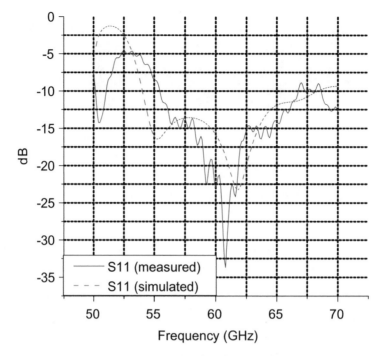

FIGURE 6.12: Measured and simulated return loss (S11) at 61.5 GHz of the series-fed 1 × 4 antenna array.

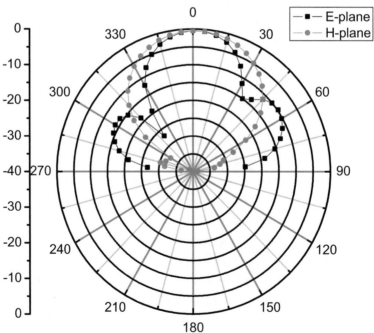

FIGURE 6.13: Simulated radiation patterns at 61.5 GHz of the series fed 1 × 4 antenna array.

CHAPTER 7

Fully Integrated Three-Dimensional Passive Front-Ends

In this chapter, two examples featuring the compact integration of antennas and filters for TDD and FDD 60 GHz applications will be used as the closing statement of the very high potential of the multilayer integration approach, especially in the mm-wave frequency range.

7.1 PASSIVE FRONT-ENDS FOR 60 GHz TIME-DIVISION DUPLEXING (TDD) APPLICATIONS

In the 60 GHz front-end module development, the compact and efficient integration of the antenna and filter is a crucial issue in terms of real estate efficiency and performance improvement in terms of high level of band selectivity, reduced parasitic problems, and low filtering loss. Particularly, when the integration is constructed in a high-ε_r material, such as low temperature cofired ceramic (LTCC), the excitation of strong surface waves causes unwanted coupling between the antenna and the rest of the components on the board. Using quasi-elliptic filters and the series-fed array antenna, it is now possible to realize a V-band compact integrated front-end.

7.1.1 Topologies

The three-dimensional (3D) overview and cross-sectional view of the topology chosen as the benchmark for the efficient compact integration are shown in Fig. 7.1(a) and (b) respectively. The four-pole quasielliptic filter and the 1×4 series fed array antenna are located on the top metallization layer [metal1 in Fig. 7.1(b)] and are connected together with a tapered microstrip transition [61] as shown in Fig. 7.1(a). The design of the tapered microstrip transition aims to annihilate the parasitic modes from the 50 Ω microstrip lines discontinuities between the two devices and to maintain a good impedance matching (20 dB bandwidth $\approx$10%). The ground planes of the filter and the antenna are located on metal 3 and 7, respectively. The ground plane of the filter is terminated with a 175 μm extra metal pad from the edge of the antenna feedline due to LTCC design rules, and the two ground planes on metal 3 and 7 are connected together with a via array as presented in Fig. 7.1.

The fabricated integrated front-end occupies an area of $9.616 \times 1.542 \times 0.318$ mm^3 including the CPW measurement pads.

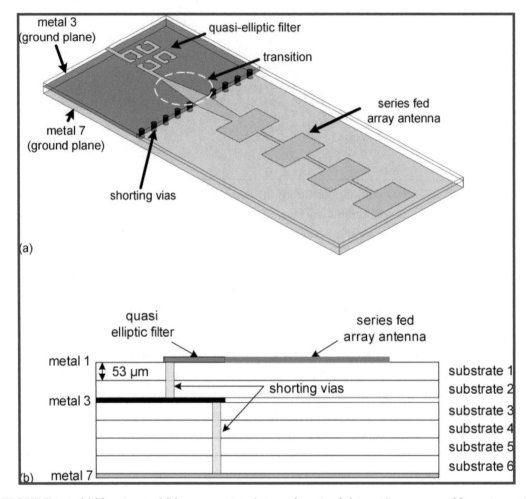

FIGURE 7.1: (a) Top view and (b) cross-sectional view of a series fed 1 × 4 linear array of four microstrip patches. All dimensions indicated in (a) are in mm.

7.1.2   Performance Discussion

Figure 7.2 shows the simulated and measured return losses of the integrated structure. It can be observed that the 10-dB return loss bandwidth is approximately 4.8 GHz (59.2–64 GHz) that is slightly wider than the simulation of 4 GHz (60–64 GHz). The slightly increased bandwidth may be attributed to the parasitic radiation from the feedlines and from the transition as well as from the edge effects of the discontinuous ground plane.

7.2   PASSIVE FRONT-ENDS FOR 60 GHz FREQUENCY-DIVISION DUPLEXING APPLICATIONS

The optimal integration of antennas and duplexers into 3D 59–64 GHz frequency-division duplexing (FDD) transceiver modules is highly desirable since it not only reduces cost, size, and system

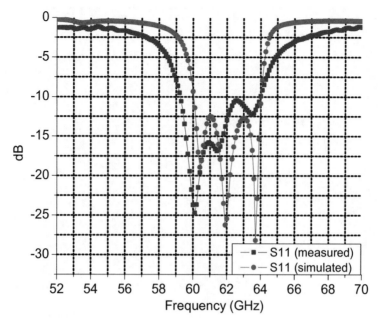

FIGURE 7.2: Comparison between measured and simulated return loss (S11) of the integrated filter and antenna functions.

complexity but also achieves a high level of band selectivity and spurious suppression, providing a high level of isolation between two channels. Although cost, electrical performance, integration density, and packaging capability are often at odds in radio frequency (RF) front-end designs, the performance of the module can be significantly improved by employing the 3D integration of filters and antennas using the flexibility of multilayer architecture on LTCC.

In this section, the full integration of the two Rx and Tx filters and the dual-polarized cross-shaped antenna that covers both Rx (1st) and Tx (2nd) channels are proposed employing the presented designs of the filters. The filters' matching (>10 dB) toward the antenna and the isolation (>45 dB) between Rx and Tx paths comprise the excellent features of this compact 3D design. The stringent demand of high isolation between two channels induces the advanced design of a duplexer and an antenna as a fully integrated function for V-band front-end module.

7.2.1   Topologies

The 3D overview and the cross-sectional view of the topology chosen for the integration are shown in Fig. 7.3(a) and (b), respectively. A cross-shaped patch antenna designed in Section 6.3 to cover two bands between 59–64 GHz (1st channel: 59–61.5 GHz, 2nd channel: 61.75–64 GHz) is located at the lowest metal layer [M11 in Fig. 7.3(b)]. The cross-shaped geometry was utilized to decrease the cross-polarization, which could potentially contribute to unwanted side lobes in

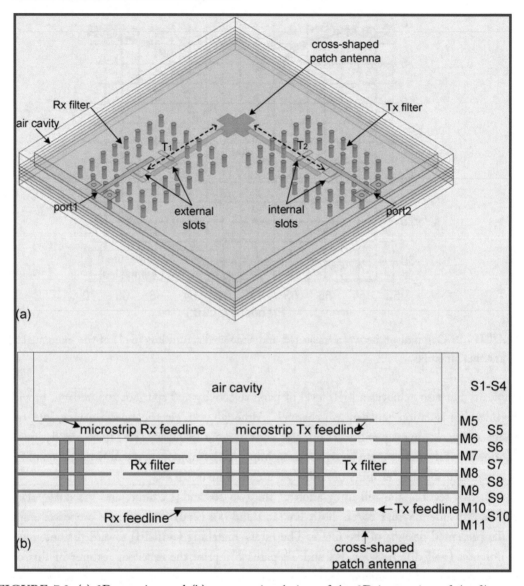

FIGURE 7.3: (a) 3D overview and (b) cross-sectional view of the 3D integration of the filters and antennas using LTCC multilayer technologies.

the radiation pattern. The cross-channel isolation can be improved by receiving and transmitting signals in two orthogonal polarizations.

The feedlines and the patch are implemented into different vertical metal layers (M10 and M11, respectively), and then the end-gap capacitive coupling is realized by overlapping the end of the embedded microstrip feedlines and the patch. The overlap distance for Rx and Tx feedline is approximately 0.029 and 0.03 mm, respectively. The common ground plane for the feedlines and the patch is placed one layer above the feedlines as shown in Fig. 7.3(b).

The two antenna feedlines [Rx feedline and Tx feedline in Fig. 7.3(b)] are commonly utilized as the filters' feedlines that excite the Rx and Tx filters accordingly through external slots placed at M9 in Fig. 7.3(b). The lengths of Rx and Tx feedlines [T_1 and T_2 in Fig. 7.3(a)] connecting the cross-shaped antenna to the Rx and Tx filters, respectively, are initially set up to be one guided wavelength at the corresponding center frequency of each channel and are optimized using high-frequency structure simulator (HFSS) simulator in the way discussed in Section 5.4.3 (T_1: 2.745 mm, T_2: 2.650 mm). The 3D Rx and Tx filters (see Fig. 33) designed in Section 5.3.2 are directly integrated to the antenna, exploiting the design parameters listed in Table 5.2. The integrated filters and antenna function occupies six substrate layers (S5–S10: 600 μm). The remaining four substrate layers [S1–S4 in Fig. 7.3(b)] are dedicated to the air cavities reserved for burying RF active devices [RF receiver and transmitter monolithic-microwave integrated circuits (MMICs)] that are located beneath the antenna on purpose not to interfere with the antenna performance and to be highly integrated with the microstrip (Rx/Tx) feedlines, leading to significant volume reduction, as shown in Fig. 7.3. The cavities are fabricated removing the inner portion of the LTCC material outlined by the successively punched vias. The deformation factor of a cavity that is defined to be the physical depth difference between the designed one and the fabricated one is stable in LTCC process when the depth of the

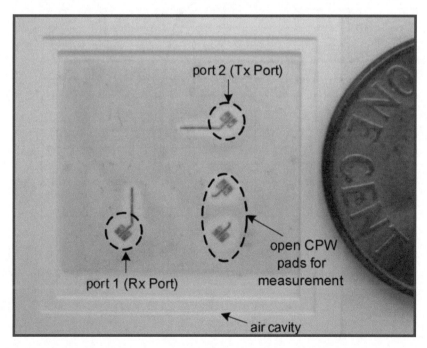

FIGURE 7.4: Photograph of the top view of the integrated function of Rx/Tx cavity filters and cross-shaped patch antenna with the air cavity top.

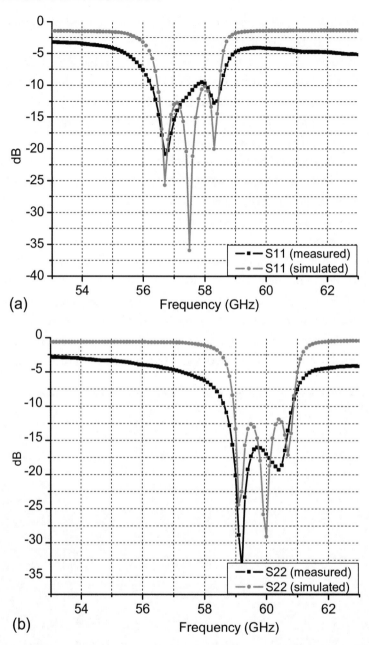

FIGURE 7.5: Comparison between measured and simulated return loss (a) S11 of the 1st channel (b) S22 of the 2nd channel.

cavity is less than two-thirds of the height of the board. Since we have chosen the air cavity depth of 400 μm, which is suitable for Rx/Tx MMIC chipsets, to enable the full integration of MMICs and passive front-end components, we can minimize the fabrication tolerances effect of an air cavity to the other integrated circuitries. Figure 7.4 shows the photograph of the integrated device that is equipped with one air cavity at the top layers. The device occupies an area of $7.94 \times 7.82 \times 1\,\text{mm}^3$ including the CPW measurement pads.

7.2.2 Performance Discussion

Figure 7.5 shows the simulated and measured return losses (S11/S22) of the integrated structure. In the simulation, the higher dielectric constant ($\varepsilon_r = 5.5$) and 5% increase in the volume of cavity were applied. It is observed from the 1st channel that the 10-dB return loss bandwidth is approximately 2.4 GHz ($\sim$4.18%) at the center frequency of 57.45 GHz that is slightly wider than the simulation of 2.1 GHz ($\sim$3.65%) at 57.5 GHz as shown in Fig. 7.5(a). The slightly increased bandwidth may be attributed to parasitic radiation from the feedlines or the measurement pads. In Fig. 7.5(b), the return loss measurement from the 2nd channel exhibits also a wider bandwidth of 2.3 GHz ($\sim$3.84%) at the center frequency of 59.85 GHz compared to the simulated value of 2.1 GHz ($\sim$3.51%) at that of 59.9 GHz. The measured channel-to-channel isolation is illustrated in Fig. 7.6. The measured isolation is better than 49.1 dB across the 1st band (56.2–58.6 GHz) and better than 51.9 dB across the 2nd band (58.4–60.7 GHz).

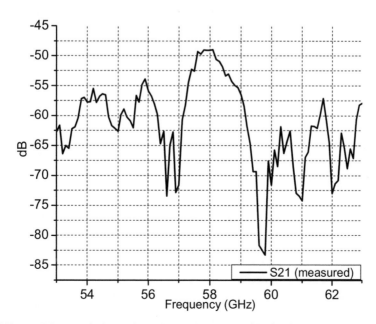

FIGURE 7.6: Measured channel-to-channel isolation (S21) of the integrated structure.

REFERENCES

[1] K. Lim, S. Pinel, M. F. Davis, A. Sutono, C. -H. Lee, D. Heo, A. Obatoynbo, J. Laskar, E. M. Tentzeris, and R. Tummala, "RF-system-on-package (SOP) for wireless communications," *IEEE Microwave Magazine*, vol. 3, no. 1, pp. 88–99, Mar. 2002, doi:10.1109/MMW.2002.990700.

[2] C. H. Doan, S. Emami, D. A. Sobel, A. M. Niknejad, and R. W. Brodersen, "Design considerations for 60 GHz CMOS radios," *IEEE Communication Magazine*, vol. 42, no. 12, pp. 132–140, Dec. 2004, doi:10.1109/MCOM.2004.1367565.

[3] H. H. Meinel, "Commercial applications of millimeter waves History, Present Status, and Future Trends," *IEEE Transaction on Microwave Theory and Technique*, vol. 43, no. 7, pp. 1639–1653, July 1995, doi:10.1109/22.392935.

[4] M. M. Tentzeris, J. Laskar, J. Papapolymerou, D. Thompson, S. Pinel, R. L. Li, J. H. Lee, G. DeJean, S. Sarkar, R. Pratap, R. Bairavasubramanian, and N. Papageorgiou, "RF SOP for multi-band RF and millimeter-wave systems," *Advanced Packaging Magazine*, pp. 15–16, April 2004.

[5] D. C. Thompson, O. Tantot, H. Jallageas, G. E. Ponchak, M. M. Tentzeris, and J. Papapolymerou, "Characterization of Liquid Crystal Polymer (LCP) Material and Transmission Lines on LCP Substrates from 30–110 GHz," *IEEE Transactions on Microwave Theory and Techniques*, vol.52, no. 4, pp. 1343–1352, April 2004, doi:10.1109/TED.2003.810465.

[6] J. Lee, K. Lim, S. Pinel, G. DeJean, R. L. Li, C. -H. Lee, M. F. Davis, M. Tentzeris, and J. Laskar, "Advanced System-on-Package (SOP) Multilayer Architectures for RF/Wireless Systems up to Millimeter-Wave Frequency Bands," in *Proc. Asian Pacific Microwave Conference*, Seoul, Korea, Nov. 2003, pp. FA5_01.

[7] V. Kondratyev, M. Lahti, and T. Jaakola, "On the design of LTCC filter for millimeter-waves," in *2003 IEEE MTT-S Int. Microwave Sym. Dig.*, Philadelphia, PA., June 2003, pp. 1771–1773.

[8] R. L. Li, G. DeJean, M. M. Tentzeris, J. Laskar, and J. Papapolymerou, "LTCC Multilayer based CP Patch Antenna Surrounded by a Soft-and-Hard Surface for GPS Applications," in *2003 IEEE-APS Symposium*, Columbus, OH, June 2003, pp. II.651–654, doi:10.1109/IEDM.1989.74183.

[9] C. H. Lee, A. Sutono, S. Han, K. Lim, S. Pinel, J. Laskar, and E. M. Tentzeris, "A Compact LTCC-based Ku-band Transmitter Module," *IEEE Transactions on Advanced Packaging*, Vol. 25, No. 3, Aug. 2002, pp. 374–384, doi:10.1109/IEDM.1990.237127.

[10] Y. Rong, K. A. Zaki, M. Hageman, D. Stevens, and J. Gipprich, "Low-Temperature Cofired Ceramic (LTCC) Ridge Waveguide Bandpass Chip Filters," *IEEE Transaction on Microwave Theory and Technique*, Vol. 47, No. 12, pp. 2317–2324, Dec. 1999, doi:10.1109/IEDM.2000.904284.

[11] Yong Huang, "A Broad-Band LTCC Integrated Transition of Laminated Waveguide to Air-Filled Waveguide for Millimeter-Wave Applications," *IEEE Transaction on Microwave Theory and Technique*, Vol. 51, No. 5, pp. 1613–1617, May 2003, doi:10.1109/96.544361.

[12] W. -Y. Leung, K. -K. M. Cheng, and K. -L. Wu, "Multilayer LTCC Bandpass Filter Design with Enhanced Stopband Characteristics," *IEEE Microwave and Wireless Components Letters*, Vol. 12, No. 7, pp. 240–242, May 2002.

[13] Y. Rong, K. A. Zaki, M. Hageman, D. Stevens, and J. Gipprich, "Low Temperature Cofired Ceramic (LTCC) Ridge Waveguide Multiplexers," in *2000 IEEE MTT-S Int. Microwave Sym. Dig.*, Boston, MA., June 2000, pp. 1169–1172, doi:10.1109/TMTT.2004.825738.

[14] R. Lucero, W. Qutteneh, A. Pavio, D. Meyers, and J. Estes, "Design of An LTCC Switch Diplexer Front-End Module for GSM/DCS/PCS Application," in *2001 IEEE Radio Frequency Integrated Circuit Sym.*, Phoenix, AZ., May 2001, pp. 213–216.

[15] C. H. Lee, A. Sutono, S. Han, K. Lim, S. Pinel, J. Laskar, and E. M. Tentzeris, "A Compact LTCC-based Ku-band Transmitter Module," *IEEE Transactions on Advanced Packaging*, Vol. 25, No. 3, pp. 374–384, Aug. 2002.

[16] B. G. Choi, M. G. Stubbs, and C. S. Park, "A Ka-Band Narrow Bandpass Filter Using LTCC Technology," *IEEE Microwave and Wireless Components Letters*, Vol. 13, No. 9, pp. 388–389, Sep. 2003.

[17] V. Piatnitsa, E. Jakku, and S. Leppaevuori, "Design of a 2-Pole LTCC Filters for Wireless Communications," *IEEE Transactions on Wireless Communications*, Vol. 3, No. 2, Mar. 2004, pp. 379–381, doi:10.1109/TADVP.2002.805315.

[18] M. J. Hill, R. W. Ziolkowski, and J. Papapolymerou, "Simulated and Measured Results from a Duroid-Based Planar MBG Cavity Resonator Filter," *IEEE Microwave and Wireless Components Letters*, Vol. 10, No. 12, pp. 528–530, Dec. 2000, doi:10.1109/22.808977.

[19] H. -J. Hsu, M. J. Hill, J. Papapolymerou, and R. W. Ziolkowski, "A Planar X-Band Electromagnetics Band-Gap (EBG) 3-Pole Filter," *IEEE Microwave and Wireless Components Letters*, Vol. 12, No. 7, pp. 255–257, July 2002, doi:10.1109/TMTT.2003.810146.

[20] C. A. Tavernier, R. M. Henderson, and J. Papapolymerou, "A Reduced-Size Silicon Micromachined High-Q Resonator at 5.7 GHz," *IEEE Transaction on Microwave Theory and Technique*, Vol. 50, NO. 10, pp. 2305–2314, Oct. 2002, doi:10.1109/LMWC.2002.801130.

[21] A. El-Tager, J. Bray, and L. Roy, "High-Q LTCC Resonators For Millimeter Wave Applications," in *2003 IEEE MTT-S Int. Microwave Sym. Dig*, Philadelphia, PA., June 2003, pp. 2257–2260.

[22] P. Ferrand, D. Baillargeat, S. Verdeyme, J. Puech, M. Lahti, and T. Jaakola, "LTCC reduced-size bandpass filters based on capacitively loaded cavities for Q band application," in *2005 IEEE MTT-S Int. Microwave Sym. Dig*, Long Beach, CA., June 2005, pp. 1789–1792.

[23] X. Gong, W. J. Chappell, and L. P. B. Katehi, "Multifunctional Substrates For High-Frequency Applications," *IEEE Microwave and Wireless Components Letters*, Vol. 13, No. 10, pp. 428–430, Oct. 2003, doi:10.1109/TADVP.2002.805315.

[24] Y. C. Lee, W. -I. Chang, Y. H. Cho, and C. S. Park, "A Very Compact 60GHz Transmitter Integrating GaAs MMICs on LTCC Passive Circuits for Wireless Terminals Applications," in *2004 IEEE MTT-S Int. Microwave Sym. Dig*, Fort Worth, TX., Oct. 2004, pp. 313–316, doi:10.1109/LMWC.2003.817139.

[25] K. Ohata, T. Inoue, M. Funabashi, A. Inoue, Y. Takimoto, T. Kuwabara, S. Shinozaki, K. Maruhashi, K. Hosaya, and H. Nagai, "Sixty-GHz-Band Ultra-Miniature Monolithic T/R Modules for Multimedia Wireless Communication Systems," *IEEE Transaction on Microwave Theory and Technique*, vol. 44, no. 12, pp. 2354–2360, Dec. 1996, doi:10.1109/TWC.2003.821141.

[26] K. Ohata, K. Maruhashi, M. Ito, S. Kishimoto, K. Ikuina, T. Hashiguchi, K. Ikeda, and N. Takahashi, "1.25 Gbps Wireless Gigabit Ethernet Link at 60 GHz-Band," in *2003 IEEE MTT-S Int. Microwave Sym. Dig*, Philadelphia, PA, June 2003, pp. 373–376.

[27] J. Mizoe, S. Amano, T. Kuwabara, T. Kaneko, K. Wada, A. Kato, K. Sato, and M. Fujise, "Miniature 60 GHz Transmetter/Receiver Modules on AIN Multi-Layer High Temperature Co-Fired Ceramic," in *1999 IEEE MTT-S Int. Microwave Sym. Dig*, Anaheim, CA, June 1999, pp. 475–478, doi:10.1109/LMWC.2002.801136.

[28] A. Tavakoli, N. Darmvandi, and R. M. Mazandaran, "Analysis of cross-shaped dual-polarized microstrip patch antennas," in *1995 IEEE AP-S Int. Sym. Dig.*, Newport Beach, CA., June 1995, pp. 994–997, doi:10.1109/TMTT.2002.80342.

[29] M. F. Davis, A. Sutono, A. Obatoyinbo, S. Chakraborty, K. Lim, S. Pinel, J. Laksar, and R. Tummala, "Integrated RF architectures in fully-organic SOP technology," in *Proc. 2001 IEEE EPEP Topical Meeting*, Boston, MA, Oct. 2001, pp. 93–96.

[30] K. Lim, A. Obatoyinbo, M. F. Davis, J. Laksar, and R. Tummala, "Development of planar antennas in multi-layer package for RF-system-on-package applications," in *Proc. 2001 IEEE EPEP Topical Meeting*, Boston, MA, Oct. 2001, pp. 101–104.

[31] M. F. Davis, A. Sutono, K. Lim, J. Laksar, V. Sundaram, J. Hobbs, G. E. White, and R. Tummala, "RF-microwave multi-layer integrated passives using fully organic system-on-package (SOP) technology," in *Proc. 2001 IEEE International Microwave Symposium*, vol. 3, Phoenix, AZ, May 2001, pp. 1731–1734, doi:10.1109/LMWC.2003.818525.

[32] J. M. Hobbs, S. Dalmia, V. Sundaram, L. Wan, W. Kim, G. White, M. Swaminathan, and R. Tummala, "Development and characterization of embedded thin-film capacitors for mixed

signal applications on fully organic system-on-package technology," in *Proc. IEEE 2002 Radio and Wireless Conference, RAWCON 2002*, Boston, MA, Aug. 2002, pp. 201–204.

[33]   R. Ulrich and L. Schaper, Eds., *Integrated Passive Component Technology*: IEEE Press/Wiley, 2003.

[34]   J. Laksar, M. Tentzeris, K. Lim, S. Pinel, M. Davis, A. Rhagavan, M. Maeng, S. -W. Yoon, and R. Tummala, "Advanced system-on-package RF front-ends for emerging wireless communications," in *Proc. 2002 Asian-Pacific Microwave Symposium*, Kyoto, Japan, Nov. 2002, pp. III. 1703–1708.

[35]   M. M, Tentzeris, J. Laskar, J. Papapolymerou, S. Pinel, V. Palazzari, R. Li, G. DeJean, N. Papageorgiou, D. Thompson, R. Bairavasubramanian, S. Sarkar, and J. -H. Lee, "3-D-integrated RF and millimeter-wave functions and modules using liquid crystal polymer (LCP) system-on-package technology," *IEEE Transaction on Advanced Packaging*, vol. 27, no. 2, pp. 332–340, May 2004.

[36]   K. Lim, A. Obatoyinbo, A. Sutuno, S. Chakraborty, C. Lee, E. Gebara, A. Raghavan, and J. Laskar, "A highly integrated transceiver module for 5.8 GHz OFDM communication system using multi-layer packaging technology," in *IEEE MTT-S Int. Microwave Symp, Dig.*, vol. 1, 2001, pp. 65–68.

[37]   W. Diels, K. Vaesen, K. Wambacq, P. Donnay, S. De Raedt, W. Engels, and M. Bolsens, "A single-package integration of RF blocks for a 5 GHz WLAN application," *IEEE Trans. Comp. Packaging. Technol. Adv. Packag.* pt. B, vol. 24, Aug. 2001, pp. 384–391, doi:10.1109/TADVP.2004.828814.

[38]   P. G. Barnwell and L. Wood, "A novel thick-film on ceramic MCM technology offering MCM-D performance," in *6th International Conf. on MCMs*, Denver, CO, 1997, pp. 48–52.

[39]   P. G. Barnwell, C. E. Free, and C. S. Aitchison, "A novel thick-film on ceramic microwave technology," in *Proc. 1998 Asian-Pacific Microwave Symposium*, Yokohama, Japan, Dec. 1998, pp. 189–192, doi:10.1109/6040.938307.

[40]   B. Geller, B. Thaler, A. Fathy, M. J. Liberatore, H. D. Chen, G. Ayers, V. Pendrick, and Y. Narayan, "LTCC-M: An enabling technology for high performance multilayer RF systems," in *IEEE MTT-S Int. Microwave Symp., Dig.*, Anaheim, CA, June 1999, pp. 189–192.

[41]   D. M. Pozar, Microwave Engineering, 2nd ed. New York: Wiley, 1998.

[42]   D. M. Pozar and D. H. Schauber, *Microstrip Antennas*, Piscataway, NJ/U.S.A.:IEEE press, 1995.

[43]   Robert E. Collin, *Foundations for Microwave Engineering*, New York, NY/U.S.A: McGraw Hill, 1992.

[44]   J. -S. Hong and M. J. Lancaster, "Coupling of microstrip square open-loop resonators for cross-coupled planar microwave filters," *IEEE Transaction on Microwave Theory and Technique*, Vol. 44, No. 12, pp. 2099–2109, Dec. 1996.

[45] J.-S. Hong and M. J. Lancaster, "Design of Highly Selective Microstrip Bandpass Filters with a Single Pair of Attenuation Poles at Finite Frequencies," *IEEE Transactions on Microwave Theory and Techniques*, vol. 48, pp. 1098–1107, July 2000.

[46] J.-G. Yook, N. I. Dib, and L. P. B. Katehi, "Characterization of High Frequency Interconnects Using Finite Difference Time Domain and Finite Element Methods," *IEEE Transactions on Microwave Theory and Techniques*, vol. 42, pp. 1727–1736, Sept. 1994, doi:10.1109/JSSC.2007.894325.

[47] Y. Cassivi and K. Wu, "Low Cost Microwave Oscillator Using Substrate Integrated Waveguide Cavity," *IEEE Microwave and Wireless Components Letters*, Vol. 13, No. 2, pp. 48–50, Feb. 2003, doi:10.1109/JSSC.2005.858626.

[48] J.-S. Hong and M. J. Lancaster, *Microstrip Filters for RF/Microwave Applications*, New York, NY/U.S.A: John Wiley $ Sons, Inc., 2001.

[49] J.-H. Lee, S. Pinel, J. Papapolymerou, J. Laskar, and M. M. Tentzeris, "Low Loss LTCC Cavity Filters Using System-on-Package Technology at 60 GHz," *IEEE Transaction on Microwave Theory and Technique*, vol. 53, no. 12, pp. 231–244, Dec. 2005, doi:10.1109/22.554558.

[50] J. Heyen, A. Gordiyenko, P. Heide, and A. F. Jacob, "Vertical Feedthroughs for Millimeter-Wave LTCC Modules," in *2003 IEEE European Microwave Conference*, Munich, Germany, Oct. 2003, pp. 411–414.

[51] H.-C. Chang and K. A. Zaki, "Evanescent-mode coupling of dual-mode rectangular waveguide filters," *IEEE Transaction on Microwave Theory and Technique*, vol. 39, no. 8, pp. 1307–1312, Aug. 1991.

[52] K. Sano and M. Miyashita, "Application of the planar I/O terminal to dual-mode dielectric-waveguide filters," *IEEE Transaction on Microwave Theory and Technique*, vol. 48, no. 12, pp. 2491–2495, Dec. 2000, doi:10.1109/TADVP.2004.831868.

[53] A. I. Atia and A. E. Williams, "Narrow-bandpass waveguide filters," *IEEE Transaction on Microwave Theory and Technique*, vol. MTT-20, no. 4, pp. 258–265, April 1972, doi:10.1109/TMTT.2004.825738.

[54] A. I. Atia and A. E. Williams, "Nonminimum-phase optimum-amplitude bandapss waveguide filters," *IEEE Transaction on Microwave Theory and Technique*, vol. MTT-22, no. 4, pp. 425–431, April 1974.

[55] D. Deslandes and K. Wu, "Substrate Integrated Waveguide Dual-Mode Filters for Broadband Wireless Systems," in *2003 Radio and Wireless Conf.*, Boston, MA, Aug. 2003, pp. 385–388.

[56] M. Guglielmi, P. Jarry, E. Kerherve, O. Roquerbrun, and D. Schmitt "A new family of all-inductive dual-mode filters," IEEE Transaction on Microwave Theory and Technique, vol. 49, no. 10, pp. 1764–1769, Oct. 2001, doi:10.1109/8.410216.

[57] P. Savi, D. Trinchero, R. Tascone, and R. Orta "A new approach to the design of dual-mode rectangular waveguide filters with distributed coupling," *IEEE Transaction on Microwave Theory and Technique*, vol. 45, no. 2, pp. 221–228, Feb. 1997.

[58] J. -F. Liang, X. -P. Liang, K. A. Zaki, and A. E. Atia "Dual-mode dielectric or air-filled rectangular waveguide filters," *IEEE Transaction on Microwave Theory and Technique*, vol. 42, no. 7, pp. 1330–1336, July. 1994.

[59] A. E. Williams and A. E. Atia, "Dual-mode canonical waveguide filters," *IEEE Transaction on Microwave Theory and Technique*, vol. MTT-25, no. 12, pp. 1021–1026, Dec 1977, doi:10.1109/LMWC.2006.877130.

[60] C. Kdsia, R. Cameron, and W. -C. Tang, "Innovations in microwave filters and multiplexing networks for communications satellite systems," *IEEE Transaction on Microwave Theory and Technique*, vol. 40, no. 6, pp. 1133–1149, June 1992.

[61] L. Accatino, G. Bertin, and M. Mongiardo, "Elliptical cavity resonators for dual-mode narrow-band filters," *IEEE Transaction on Microwave Theory and Technique*, vol. 45, no. 12, pp. 2393–2401, Dec. 1997.

[62] A. E. Williams, "A Four-Cavity Elliptic Waveguide Filter," *IEEE Transaction on Microwave Theory and Technique*, vol. MTT-18, no. 12, pp. 1109–1114, Dec. 1970.

[63] I. Awai, A. C. Kundu, and T. Yamashita, "Equivalent-circuit representation and explanation of attenuation poles of a dual-mode dielectric-resonator bandpass filter," *IEEE Transaction on Microwave Theory and Technique*, vol. 46, no. 11, pp. 2159–2163, Dec. 1998, doi:10.1109/22.543968.

[64] M. Sagawa, K. Takahashi, and M. Makimoto, "Miniaturized hairpin resonator filters and their application to receiver front-end MIC's," *IEEE Transaction on Microwave Theory and Technique*, vol. 37, no. 12, pp. 1991–1997, May 1989, doi:10.1109/22.848492.

[65] K. A. Zaki, C. Chen, and A. E. Atia, "A circuit model of probes in dual-mode cavities," *IEEE Trans. Microwave Theory Tech.*, vol. 36, pp. 1740–1746, Dec. 1988, doi:10.1109/22.310581.

[66] J. B. Thomas, "Cross-coupling in coaxial cavity filters – a tutorial overview," *IEEE Transaction on Microwave Theory and Technique*, vol. 51, no. 4, pp. 1368–1376, April 2003, doi:10.1109/LMWC.2003.808720.

[67] F. Purroy and L. Pradell, "New theoretical analysis of the LRRM calibration technique for vector network analyzers," *IEEE Transaction on Instrumentation and Measurement*, vol. 50, issues 5, pp. 1307–1314, Oct. 2001.

[68] M. J. Vaughan, K. Y. Hur, and R. C. Compton, "Improvement of microstrip patch antenna radiation patterns," *IEEE Trans. Antennas Propagat.*, vol. 42, no. 6, pp. 882–885, June 1994.

[69] R. Gonzalo, P. de. Maagt, and M. Sorolla, "Enhanced patch-antenna performance by suppressing surface waves using photonic-bandgap substrates," *IEEE Trans. Microwave Theory Tech.*, vol. 47, no. 11, pp. 2131–2138, Nov. 1999.

[70] R. Coccioli, F. -R. Yang, K. -P. Ma, and T. Itoh, "Aperture-coupled patch antenna on UC-PBG substrate," *IEEE Trans. Microwave Theory Tech.*, vol. 47, no. 11, pp. 2123–2130, Nov. 1999, doi:10.1109/22.85405.

[71] R. L. Li, G. DeJean, J. Papapolymerou, J. Laskar, and M. M. Tentzeris, "Radiation-pattern improvement of patch antennas on a large-size substrate using a compact soft surface structure and its realization on LTCC multilayer technology," *IEEE Trans. Antennas and Propagation*, vol. 53, no. 1, pp. 200–208, Jan. 2000, doi:10.1109/22.899003.

[72] M. Tentzeris, R. L. Li, K. Lim, M. Maeng, E. Tsai, G. DeJean, and J. Laskar, "Design of compact stacked-patch antennas on LTCC technology for wireless communication applications," *Proceedings of IEEE Antenna and Propagation Society International Symposium*, vol. 2, pp. 500–503, June 2002, doi:10.1109/TMTT.1972.1127732.

[73] C. Balanis, *Antenna Theory*. Canada: John Wiley & Sons, Inc, 1997, doi:10.1109/TMTT.1974.1128242.

[74] P. F. M. Smulder, "Exploiting the 60 GHz Band for Local Wireless Multimedia Access: Prospects and Future Directions," *IEEE Communications Magazine*, vol.40, pp. 140–147, Jan. 2002.

[75] L. Xue, C. C. Liu, H. -S Kim, and S. Tiwari, "Three-dimensional integration: technology, use, and issues for mixed-signal applications," *IEEE Trans. on Electron Devices.*, vol. 50, no. 3, pp. 601–608, Mar. 2003, doi:10.1109/22.954782.

[76] S. Kawamura, N. Sasaki, I. Iwai, M. Nakano, and M. Takagi, "Three-dimensional CMOS IC's fabricated by using beam recrystallization," *IEEE Electron Device Lett.*, vol. EDL-4, pp. 601–608, Mar. 2003, doi:10.1109/22.557603.

[77] Y. Akasaka and T. Nishimura, "Concept and basic technologies for 3-D IC sturcture," in *IEDM Tech. Dig.*, 1986, pp. 488–491, doi:10.1109/22.299726.

[78] T. Kunio, K. Oyama, Y. Hayashi, and M. Morimoto, "Three-dimensional IC's, having four stacked active device layers," in *IEDM Tech. Dig.*, 1989, pp. 837–840, doi:10.1109/TMTT.1977.1129267.

[79] K. Yamazaki, Y. Itoh, A. Wada, K. Morimoto, and Y. Tomita, "4-layer 3-D IC technologies for parallel signal processing," in *IEDM Tech. Dig.*, 1990, pp. 599–602, doi:10.1109/22.141345.

[80] K. W. Lee, T. Nakamura, T. Ono, Y. Yamada, T. Mizukusa, H. Hashimoto, K. T. Park, H. Kurino, and M. Koyanagi, "Three-dimensional shared memory fabricated using wafer stacking technology," in *IEDM Tech. Dig.*, 2000, pp. 165–168, doi:10.1109/22.643850.

[81] M. B. Kleiner, S. A. Kuhn, P. Ramm, and W. Weber, "Performance improvement of the memory hierarchy of RISC-systems by application of 3-D technology," *IEEE Trans. on Comp. Packag. Manufact. Technol.*, vol. 19, pp. 709–718, Nov. 1996, doi:10.1109/TMTT.1970.1127419.

[82] R. L. V. Tuyl, "Unlicensed millimeter wave communications: A new opportunity for MMIC technology at 60 GHz," in *IEEE GaAs IC Symp.*, 1996, pp. 3–5, doi:10.1109/22.739300.

[83] H. Daembkes, "GaAs MMIC based components and frontends for millimeter-wave communication and sensor system," in *Microwave Sys. Conf.*, 1995, pp. 83–86, doi:10.1109/22.44113.

[84] S. Reynolds, "60 GHz transceiver circuits in SiGe Bipolar Technology," in *IEEE Int'l Solid-State Circuits Conf. Dig. Tech. Papers*, Feb. 2004, pp. 442–443, doi:10.1109/22.17408.

[85] T. Yao, M. Q. Gordon, K. K. W. Tang, K. H. K. Yau, M. -T. Yang, P. Schvan, and S. P. Voinigescu, "Algorithmic design of CMOS LNAs and PAs for 60-GHz radio," IEEE Journal of Solid-State Circuit, vol. 42, no. 5, pp. 1044–1057, May 2007, doi:10.1109/TMTT.2003.809180.

[86] B. Razavi, "A 60-GHz CMOS receiver front-end," *IEEE Journal of Solid-State Circuit,* vol. 41, no. 1, pp. 17–22, Jan. 2006, doi:10.1109/19.963202.

[87] V. A. Chiriac and T. -Y. T. Lee, "Thermal assessment of RF-integrated LTCC front end modules," *IEEE Trans. on Adv. Packag.*, vol. 27, no. 3, pp. 545–557, Aug. 2004, doi:10.1109/8.301717.

[88] D. C. Thompson, O. Tantot, H. Jallageas, G. E. Ponchak, and M. M. Tentzeris, "Characterization of liquid crystal polymer (LCP) material and transmission lines on LCP substrates from 30–110 GHz," *IEEE Trans. Microwave Theory Tech.*, vol. 52, no. 4, pp. 1343–1352, Apr. 2004, doi:10.1109/22.798009.

[89] K. Jayaraj, T. E. Noll, and D. R. Singh, "RF characterization of a low-cost multichip packaging technology for monolithic microwave and millimeter wave integrated circuits," *in URSI Int. Signals, Systems, and Electronics Symp.*, Oct. 1995, pp. 443–446, doi:10.1109/22.798008.

[90] E. C. Culberston, "A new laminate material for high performance PCBs: Liquid crystal polymer copper clad films," in *IEEE Electronic Components and Technology Conf.*, May 2002, pp. 520–523, doi:10.1109/TAP.2004.840754.

[91] K. Jayaraj, T. E. Noll, and D. R. Singh, "A low cost multichip packaging technology for monolithic microwave integrated circuits," *IEEE Trans. Antennas Propagat.*, vol. 43, pp. 992–997, Sept. 1995.

[92] B. Farrell and M. St. Lawrence, "The processing of liquid crystalline polymer printed circuits," in *IEEE Electronic Components and Technology Conf.*, May 2002, pp. 667–671.

[93] C. Murphy, private communication, Jan. 2004.

[94] D. C. Thompson, M. M. Tentzeris, and J. Papapolymerou, "Packaging of MMICs in multi-layer LCP substrates," *IEEE Microwave Wireless Compon. Lett.*, vol. 16, iss. 7, pp. 410–412, July 2006, doi:10.1109/35.978061.

[95] Y. Li and M. M. Tentzeris, "Design and characterization of novel paper-based inkjet-printed RFID and microwave structures for telecommunication and sensing applications," in *2007 IEEE MTT-S Int. Microwave Sym. Dig.*, Honolulu, HW, June 2007, pp. 1633–1636.

Author Biography

Manos M. Tentzeris was born and grew up in Piraeus, Greece. He graduated from Ionidios Model School of Piraeus in 1987 and he received the Diploma degree in Electrical Engineering and Computer Science (Magna Cum Laude) from the National Technical University in Athens, Greece, in 1992 and the M.S. and Ph.D. degrees in Electrical Engineering and Computer Science from the University of Michigan, Ann Arbor in 1993 and 1998.

He is currently an Associate Professor with the School of ECE, Georgia Tech and he has published more than 250 papers in refereed Journals and Conference Proceedings, 1 book and 8 book chapters, while he is in the process of writing 2 books. He is currently the Georgia Electronic Design Center Associate Director for RFID/Sensors research, while he had been the GT-Packaging Research Center (NSF-ERC) Associate Director for RF research and the leader of the RF/Wireless Packaging Alliance from 2003-2006. Also, Dr. Tentzeris is the Head of the A.T.H.E.N.A. Research Group (20 students and researchers) and has established academic programs in Highly Integrated/Multilayer Packaging for RF and Wireless Applications using ceramic and organic flexible materials, paper-based RFID's and sensors, Microwave MEM's, SOP-integrated (UWB, mutliband, conformal) antennas and Adaptive Numerical Electromagnetics (FDTD, MultiResolution Algorithms). He was the 1999 Technical Program Co-Chair of the 54th ARFTG Conference and he is currently a member of the technical program committees of IEEE-IMS, IEEE-AP and IEEE-ECTC Symposia. He will be the TPC Chair for the IMS 2008 Conference. He was the Chairman for the 2005 IEEE CEM-TD Workshop. He was the Chair of IEEE-CPMT TC16 (RF Subcommittee) and he was the Chair of IEEE MTT/AP Atlanta Sections for 2003. He is a Senior Member of IEEE, a member of MTT-15 Committee, an Associate Member of European Microwave Association (EuMA), a Fellow of the Electromagnetics Academy, and a member of Commission D, URSI and of the Technical Chamber of Greece. His hobbies include basketball, ping-pong and travel.

Jong-Hoon Lee received the B.S. Degree in electrical engineering from The Pennsylvania State University, University Park, with high honor in the spring of 2001. He joined the electrical engineering at The Georgia Institute of Technology, in the fall of 2001 and received an M.S. degree in the fall of 2004 and a Ph.D. in the summer of 2007 under the advice of Prof. Manos M. Tentzeris. Dr. Lee is now the senior CAE engineer at RFMD in the design integration department.

His research interests are SOP and SIP packaging technologies for microwave/mmW systems, passive/active circuits for RF/wireless systems, DSP-based predictors to improve the computational efficiency of the simulation of highly resonant RF geometries. Jong-Hoon Lee also researches the development of the LTCC system-on-package (SOP) module for millimeter-wave wireless systems, FDTD/Spice interface, and active devices modeling with FDTD and MRTD. He was a member of the Georgia Tech ATHENA research group, NSF packaging research center, the Georgia Electronic Design Center, and Tau Beta Pi Honor association.